Lectures on Quantum Computing, Thermodynamics and Statistical Physics

Kinki University Series on Quantum Computing

Editor-in-Chief: Mikio Nakahara *(Kinki University, Japan)*

ISSN: 1793-7299

Published

Vol. 1   Mathematical Aspects of Quantum Computing 2007
edited by Mikio Nakahara, Robabeh Rahimi (Kinki Univ., Japan) &
Akira SaiToh (Osaka Univ., Japan)

Vol. 2   Molecular Realizations of Quantum Computing 2007
edited by Mikio Nakahara, Yukihiro Ota, Robabeh Rahimi, Yasushi Kondo &
Masahito Tada-Umezaki (Kinki Univ., Japan)

Vol. 3   Decoherence Suppression in Quantum Systems 2008
edited by Mikio Nakahara, Robabeh Rahimi & Akira SaiToh
(Kinki Univ., Japan)

Vol. 4   Frontiers in Quantum Information Research: Decoherence, Entanglement,
Entropy, MPS and DMRG 2009
edited by Mikio Nakahara (Kinki Univ., Japan) &
Shu Tanaka (Univ. of Tokyo, Japan)

Vol. 5   Diversities in Quantum Computation and Quantum Information
edited by edited by Mikio Nakahara, Yidun Wan & Yoshitaka Sasaki
(Kinki Univ., Japan)

Vol. 6   Quantum Information and Quantum Computing
edited by Mikio Nakahara & Yoshitaka Sasaki (Kinki Univ., Japan)

Vol. 7   Interface Between Quantum Information and Statistical Physics
edited by Mikio Nakahara (Kinki Univ., Japan) &
Shu Tanaka (Univ. of Tokyo, Japan)

Vol. 8   Lectures on Quantum Computing, Thermodynamics and Statistical Physics
edited by Mikio Nakahara (Kinki Univ., Japan) &
Shu Tanaka (Univ. of Tokyo, Japan)

Kinki University Series on Quantum Computing – Vol. 8

editors

Mikio Nakahara
Kinki University, Japan

Shu Tanaka
University of Tokyo, Japan

Lectures on Quantum Computing, Thermodynamics and Statistical Physics

World Scientific

NEW JERSEY · LONDON · SINGAPORE · BEIJING · SHANGHAI · HONG KONG · TAIPEI · CHENNAI

Published by

World Scientific Publishing Co. Pte. Ltd.

5 Toh Tuck Link, Singapore 596224

USA office: 27 Warren Street, Suite 401-402, Hackensack, NJ 07601

UK office: 57 Shelton Street, Covent Garden, London WC2H 9HE

British Library Cataloguing-in-Publication Data
A catalogue record for this book is available from the British Library.

Kinki University Series on Quantum Computing — Vol. 8
LECTURES ON QUANTUM COMPUTING, THERMODYNAMICS AND STATISTICAL PHYSICS

Copyright © 2013 by World Scientific Publishing Co. Pte. Ltd.

ISBN 978-981-4425-18-6

Printed in Singapore.

PREFACE

This volume contains lecture notes presented at the Symposium on Quantum Computing, Thermodynamics, and Statistical Physics, held from 13 to 15 March, 2012 at Kinki University, Osaka, Japan. The aim of the symposium was to exchange and share ideas among researchers working in various fields related to quantum information theory, statistical physics and thermodynamics. Lecturers were asked to make their presentations accessible to other researchers with various backgrounds. As a result, lecture notes in this volume should be accessible not only to forefront researchers but also to beginning graduate students and advanced undergraduate students. Each lecture note should also serve as an ideal textbook for one-semester graduate course.

This symposium was supported financially by "Open Research Center" Project for Private Universities: matching fund subsidy from MEXT (Ministry of Education, Culture, Sports, Science and Technology).

We would like to thank all the lecturers and participants, who made this symposium invaluable. We would like to thank Shoko Kojima for her dedicated secretarial work. Finally, we would like to thank Zhang Ji and Rhaimie B Wahap of World Scientific for their excellent editorial work.

Mikio Nakahara
Shu Tanaka

Osaka and Tokyo, March 2012

CONTENTS

QUANTUM ANNEALING:
FROM VIEWPOINTS OF STATISTICAL PHYSICS,
CONDENSED MATTER PHYSICS,
AND COMPUTATIONAL PHYSICS

SHU TANAKA

Department of Chemistry, University of Tokyo,
7-3-1, Hongo, Bunkyo-ku, Tokyo, 113-0033, Japan
E-mail: shu-t@chem.s.u-tokyo.ac.jp

RYO TAMURA

Institute for Solid State Physics, University of Tokyo,
5-1-5, Kashiwanoha, Kashiwa-shi, Chiba, 277-8501, Japan

International Center for Young Scientists,
National Institute for Materials Science,
1-2-1, Sengen, Tsukuba-shi, Ibaraki, 305-0047, Japan
E-mail: tamura.ryo@nims.go.jp

In this paper, we review some features of quantum annealing and related topics from viewpoints of statistical physics, condensed matter physics, and computational physics. We can obtain a better solution of optimization problems in many cases by using the quantum annealing. Actually the efficiency of the quantum annealing has been demonstrated for problems based on statistical physics. Then the quantum annealing has been expected to be an efficient and generic solver of optimization problems. Since many implementation methods of the quantum annealing have been developed and will be proposed in the future, theoretical frameworks of wide area of science and experimental technologies will be evolved through studies of the quantum annealing.

Keywords: Quantum annealing; Quantum information; Ising model; Optimization problem

1. Introduction

Optimization problems are present almost everywhere, for example, designing of integrated circuit, staff assignment, and selection of a mode of transportation. To find the best solution of optimization problems is difficult in general. Then, it is a significant issue to propose and to develop a method for obtaining the best solution (or a better solution) of optimization problems in information science. In order to obtain the best solution, a couple of algorithms according to type of optimization problems have been formulated in information science and these methods have yielded practical applications. Furthermore, since optimization problem is to find the state where a real-valued function takes the minimum value, it can be regarded as problem to obtain the ground state of the corresponding Hamiltonian. Thus, if we can map optimization problem to well-defined Hamiltonian, we can use knowledge and methodologies of physics. Actually, in computational physics, generic and powerful algorithms which can be adopted for wide application have been proposed. One of famous methods is simulated annealing which was proposed by Kirkpatrick et al.[1,2] In the simulated annealing, we introduce a temperature (thermal fluctuation) in the considered optimization problems. We can obtain a better solution of the optimization problem by decreasing temperature gradually since thermal fluctuation effect facilitates transition between states. It is guaranteed that we can obtain the best solution definitely if we decrease temperature slow enough.[3] Then, the simulated annealing has been used in many cases because of easy implementation and guaranty.

The quantum annealing was proposed as an alternative method of the simulated annealing.[4-11] In the quantum annealing, we introduce a quantum field which is appropriate for the considered Hamiltonian. For instance, if the considered optimization problem can be mapped onto the Ising model, the simplest form of the quantum fluctuation is transverse field. In the quantum annealing, we gradually decrease quantum field (quantum fluctuation) instead of temperature (thermal fluctuation). The efficiency of the quantum annealing has been demonstrated by a number of researchers, and it has been reported that a better solution can be obtained by the quantum annealing comparison with the simulated annealing in many cases. Figure 1 shows schematic picture of the simulated annealing and the quantum annealing. In optimization problems, our target is to obtain the stable state at zero temperature and zero quantum field, which is indicated by the solid circle in Fig. 1.

Recently, methods in which we decrease temperature and quantum field simultaneously have been proposed and as a result, we can obtain a better solution than the simulated annealing and the simple quantum annealing.[12-14] Moreover, as an another example of methods in which we use both thermal fluctuation and quantum fluctuation, novel quantum anneal-

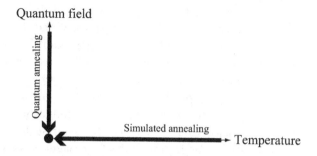

Fig. 1. Schematic picture of the simulated annealing and the quantum annealing. Our purpose is to obtain the ground state at the point indicated by the solid circle.

ing method with the Jarzynski equality[15,16] was also proposed,[17] which is based on nonequilibrium statistical physics.

In this paper, we review the quantum annealing method which is the generic and powerful tool for obtaining the best solution of optimization problems from viewpoints of statistical physics, condensed matter physics, and computational physics. The organization of this paper is as follows. In Sec. 2, we review the Ising model which is a fundamental model of magnetic systems. The realization method of the Ising model by nuclear magnetic resonance is also explained. In Sec. 3, we show a couple of implementation methods of the quantum annealing. In Sec. 4, we explain two optimization problems – traveling salesman problem and clustering problem. The quantum annealing based on the Monte Carlo method for the traveling salesman problem is also demonstrated. In Sec. 5, we review related topics of the quantum annealing – Kibble-Zurek mechanism of the Ising spin chain and order by disorder in frustrated systems. In Sec. 6, we summarize this paper briefly and give some future perspectives of the quantum annealing.

2. Ising Model

In this section we introduce the Ising model which is a fundamental model in statistical physics. A century ago, the Ising model was proposed to explain cooperative nature in strongly correlated magnetic systems from a microscopic viewpoint.[18] The Hamiltonian of the Ising model is given by

$$\mathcal{H}_{\text{Ising}} = -\sum_{i,j} J_{ij}\sigma_i^z\sigma_j^z - \sum_{i=1}^{N} h_i\sigma_i^z, \qquad \sigma_i^z = \pm 1, \qquad (1)$$

where the summation of the first term runs over all interactions on the defined graph and N represents the number of spins. If the sign of J_{ij} is positive/negative, the interaction is called ferromagnetic/antiferromagnetic interaction. Spins which are connected by ferromagnetic/antiferromagnetic interaction tend to be the same/opposite direction. The second term of the Hamiltonian denotes the site-dependent longitudinal magnetic fields. Although the Ising model is quite simple, this model exhibits inherent rich properties $e.g.$ phase transition and dynamical behavior such as melting process and slow relaxation. For instance, the ferromagnetic Ising model with homogeneous interaction ($J_{ij} = J$ for $\forall i, j$) and no external magnetic fields ($h_i = 0$ for $\forall i$) on square lattice exhibits the second-order phase transition, whereas no phase transition occurs in the Ising model on one-dimensional lattice. Onsager first succeeded to obtain explicitly free energy of the Ising model without external magnetic field on square lattice.[19] After that, a couple of calculation methods were proposed. Furthermore, these calculation methods have been improved day by day, and the new techniques which were developed in these methods have been applied for other more complicated problems. Since the Ising model is quite simple, we can easily generalize the Ising model in diverse ways such as the Blume-Capel model,[20,21] the clock model,[22,23] and the Potts model.[24,25] By analyzing these models, relation between nature of phase transition and the symmetry which breaks at the transition point has been investigated. Then, it is not too much to say that the Ising model has opened up a new horizon for statistical physics.

The Ising model can be adopted for not only magnetic systems but also systems in wide area of science such as information science. Optimization problem is one of important topics in information science. As we mention in Sec. 4, optimization problem can be mapped onto the Ising model and its generalized models in many cases. Then some methods which were developed in statistical physics often have been used for optimization problem. In Sec. 2.1, we show a couple of magnetic systems which can be well represented by the Ising model. In Sec. 2.2, we review how to create the Ising model by Nuclear Magnetic Resonance (NMR) technique as an example of experimental realization of the Ising model.

2.1. Magnetic Systems

In many cases, the Hamiltonian of magnetic systems without external magnetic field is given by

$$
\begin{aligned}
\hat{\mathcal{H}} &= -\sum_{i,j} J_{ij} \hat{\sigma}_i \cdot \hat{\sigma}_j \\
&= -\sum_{i,j} J_{ij} \left(\hat{\sigma}_i^x \cdot \hat{\sigma}_j^x + \hat{\sigma}_i^y \cdot \hat{\sigma}_j^y + \hat{\sigma}_i^z \cdot \hat{\sigma}_j^z \right),
\end{aligned} \tag{2}
$$

where $\hat{\sigma}_i^\alpha$ denotes the α-component of the Pauli matrix at the i-th site. The form of this interaction is called Heisenberg interaction. The definitions of Pauli matrices are

$$
\hat{\sigma}^x := \begin{pmatrix} 0 & 1 \\ 1 & 0 \end{pmatrix}, \qquad \hat{\sigma}^y := \begin{pmatrix} 0 & -i \\ i & 0 \end{pmatrix}, \qquad \hat{\sigma}^z := \begin{pmatrix} 1 & 0 \\ 0 & -1 \end{pmatrix}, \tag{3}
$$

where the bases are defined by

$$
|{\uparrow}\rangle := \begin{pmatrix} 1 \\ 0 \end{pmatrix}, \qquad |{\downarrow}\rangle := \begin{pmatrix} 0 \\ 1 \end{pmatrix}. \tag{4}
$$

In this case, magnetic interactions are isotropic. However, they become anisotropic depending on the surrounded ions in real magnetic materials. In general, the Hamiltonian of magnetic systems should be replaced by

$$
\hat{\mathcal{H}} = -\sum_{i,j} J_{ij} \left(c_x \hat{\sigma}_i^x \hat{\sigma}_j^x + c_y \hat{\sigma}_i^y \hat{\sigma}_j^y + c_z \hat{\sigma}_i^z \hat{\sigma}_j^z \right). \tag{5}
$$

When $|c_x|, |c_y| > |c_z|$, the xy-plane is easy-plane and the Hamiltonian becomes XY-like Hamiltonian. On the contrary, when $|c_z| > |c_x|, |c_y|$, the z-axis is easy-axis and the Hamiltonian becomes Ising-like Hamiltonian. Such anisotropy comes from crystal structure, spin-orbit coupling, and dipole-dipole coupling. Moreover, even if there is almost no anisotropy in magnetic interactions, magnetic systems can be regarded as the Ising model when the number of electrons in the magnetic ion is odd and the total spin is half-integer. In this case, doubly degenerated states exist because of the Kramers theorem. These states are called the Kramers doublet. When the energy difference between the ground states and the first-excited states ΔE is large enough, these doubly-degenerated ground states can be well represented by the $S = 1/2$ Ising spins. Table 1 shows examples of the magnetic materials which can be well represented by the Ising model on one-dimensional chain, two-dimensional square lattice, and three-dimensional cubic lattice.

Table 1. Examples of magnetic materials which can be represented by the Ising model on chain (one-dimension), square lattice (two-dimension), and cubic lattice (three-dimension).

Material	Spatial dimension	Total spin	Type of interaction	J/k_B	References
$K_3Fe(CN)_6$	One (chain)	$\frac{1}{2}$	Antiferromagnetic	−0.23 K	26–28
$CsCoCl_3$	One (chain)	$\frac{1}{2}$	Antiferromagnetic	−100 K	29,30
$Dy(C_2H_5SO_4)_2 \cdot 9\,H_2O$	One (chain)	$\frac{1}{2}$	Ferromagnetic	0.2 K	31–33
$CoCl_2 \cdot 2NC_5H_5$	One (chain)	$\frac{1}{2}$	Ferromagnetic	9.5 K	34,35
$CoCs_3Br_5$	Two (square)	$\frac{1}{2}$	Antiferromagnetic	−0.23 K	36–38
$Co(HCOO)_2 \cdot 2\,H_2O$	Two (square)	$\frac{1}{2}$	Antiferromagnetic	−4.3 K	39–42
Rb_2CoF_4	Two (square)	$\frac{1}{2}$	Antiferromagnetic	−91 K	43,44
$FeCl_2$	Two (square)	1	Ferromagnetic	3.4 K	45,46
$DyPO_4$	Three (cubic)	$\frac{1}{2}$	Antiferromagnetic	−2.5 K	47–50
$Dy_3Al_5O_{12}$	Three (cubic)	$\frac{1}{2}$	Antiferromagnetic	−1.85 K	51–53
$CoRb_3Cl_5$	Three (cubic)	$\frac{1}{2}$	Antiferromagnetic	−0.511 K	54,55
FeF_2	Three (cubic)	2	Antiferromagnetic	−2.69 K	56–59

2.2. Nuclear Magnetic Resonance

In condensed matter physics, Nuclear Magnetic Resonance (NMR) has been used for decision of the structure of organic compounds and for analysis of the state in materials by using resonance induced by electromagnetic wave. The NMR can *create* the Ising model with transverse fields, which is expected to become an element of quantum information processing. In this processing, we use molecules where the coherence times are long compared with typical gate operations. Actually a couple of molecules which have nuclear spins were used for demonstration of quantum computing.[60–75] In this section we explain how to create the Ising model by NMR.

The setup of the NMR spectrometer as a tool of quantum computing is as follows. We first put molecules which contain nuclear spins under the strong magnetic field B_0. Next we apply radio frequency $\omega^{(\text{rf})}$ magnetic field which is perpendicular to the strong magnetic field B_0. For simplicity, we here consider a molecule which contains two spins. We also assume that the considered molecule can be well described by the Heisenberg Hamiltonian. Then the Hamiltonian of this system is given by

$$\hat{\mathcal{H}} = \hat{\mathcal{H}}_{\text{mol}} + \hat{\mathcal{H}}_1^{(\text{rf})} + \hat{\mathcal{H}}_2^{(\text{rf})}, \tag{6}$$

where $\hat{\mathcal{H}}_{\text{mol}}$, $\hat{\mathcal{H}}_1^{(\text{rf})}$, and $\hat{\mathcal{H}}_2^{(\text{rf})}$ are defined by

$$\hat{\mathcal{H}}_{\text{mol}} := -h_1\hat{\sigma}_1^z - h_2\hat{\sigma}_2^z - J(\hat{\sigma}_1^x \cdot \hat{\sigma}_2^x + \hat{\sigma}_1^y \cdot \hat{\sigma}_2^y + \hat{\sigma}_1^z \cdot \hat{\sigma}_2^z), \tag{7}$$

$$\hat{\mathcal{H}}_1^{(\text{rf})} := -\Gamma_1 \cos(\omega^{(\text{rf})}t - \phi_1)(\hat{\sigma}_1^x + \gamma'\hat{\sigma}_2^x), \tag{8}$$

$$\hat{\mathcal{H}}_2^{(\text{rf})} := -\Gamma_2 \cos(\omega^{(\text{rf})}t - \phi_2)(\gamma'^{-1}\hat{\sigma}_1^x + \hat{\sigma}_2^x), \tag{9}$$

respectively. We take the natural unit in which $\hbar = 1$. The values of ϕ_1 and ϕ_2 are the phases at the time $t = 0$ of the first spin and that of the second spin, respectively. The quantities of h_i are defined by $h_i := \gamma_i B_0$, where γ_i denotes the gyromagnetic ratio of the i-th spin ($i = 1, 2$). The values of h_1 and h_2 represent energy differences between $|\uparrow\rangle$ and $|\downarrow\rangle$ of the first spin and the second spin, respectively. The coefficients Γ_1 and Γ_2 in $\hat{\mathcal{H}}_1^{(\text{rf})}$ and $\hat{\mathcal{H}}_2^{(\text{rf})}$ are the effective amplitudes of the ac magnetic field, whose definitions are $\Gamma_i := \gamma_i B_{\text{ac}}$, where B_{ac} is amplitude of the ac magnetic field. The value of γ' is defined by the ratio of the gyromagnetic ratios $\gamma' := \gamma_2/\gamma_1$.

We define the following unitary transformation:

$$\hat{U}^{(\text{R})} := \mathrm{e}^{-ih_1\hat{\sigma}_1^z t} \cdot \mathrm{e}^{-ih_2\hat{\sigma}_2^z t}. \tag{10}$$

We can change from the laboratory frame to a frame rotating with h_i around the z-axis by using the above unitary transformation. The dynamics of a density matrix can be calculated by

$$i\frac{d\hat{\rho}}{dt} = [\hat{\mathcal{H}}, \hat{\rho}]. \tag{11}$$

The density matrix on the rotating frame is given by

$$\hat{\rho}^{(R)} := \hat{U}^{(R)} \hat{\rho} \hat{U}^{(R)\dagger}. \tag{12}$$

To be the same form as Eq. (11) on the rotating frame, the Hamiltonian on the rotating frame should be

$$\hat{\mathcal{H}}^{(R)} = \hat{U}^{(R)} \hat{\mathcal{H}} \hat{U}^{(R)\dagger} - i\hat{U}^{(R)} \frac{d\hat{U}^{(R)\dagger}}{dt}. \tag{13}$$

Here we decompose the Hamiltonian on the rotating frame as

$$\hat{\mathcal{H}}^{(R)} = \hat{\mathcal{H}}_{\text{mol}}^{(R)} + \hat{\mathcal{H}}_1^{(R)(\text{rf})} + \hat{\mathcal{H}}_2^{(R)(\text{rf})}, \tag{14}$$

where the three terms are defined by

$$\hat{\mathcal{H}}_{\text{mol}}^{(R)} := \hat{U}^{(R)} \hat{\mathcal{H}}_{\text{mol}} \hat{U}^{(R)\dagger} - i\hat{U}^{(R)} \frac{d\hat{U}^{(R)\dagger}}{dt}, \tag{15}$$

$$\hat{\mathcal{H}}_1^{(R)(\text{rf})} := \hat{U}^{(R)} \hat{\mathcal{H}}_1^{(\text{rf})} \hat{U}^{(R)\dagger}, \tag{16}$$

$$\hat{\mathcal{H}}_2^{(R)(\text{rf})} := \hat{U}^{(R)} \hat{\mathcal{H}}_2^{(\text{rf})} \hat{U}^{(R)\dagger}. \tag{17}$$

The intramolecular magnetic interaction Hamiltonian on the rotating frame $\hat{\mathcal{H}}_{\text{mol}}^{(R)}$ can be calculated as

$$\hat{\mathcal{H}}_{\text{mol}}^{(R)} = J \begin{pmatrix} 0 & 0 & 0 & 0 \\ 0 & 0 & e^{i(h_2-h_1)t} & 0 \\ 0 & e^{-i(h_2-h_1)t} & 0 & 0 \\ 0 & 0 & 0 & 0 \end{pmatrix} - J\hat{\sigma}_1^z \hat{\sigma}_2^z \simeq -J\hat{\sigma}_1^z \hat{\sigma}_2^z. \tag{18}$$

The approximation is valid when $|h_2 - h_1|\tau \gg 1$, where τ is a characteristic time scale since the exponential terms are averaged to vanish. The radio frequency magnetic field Hamiltonian on the rotating frame $\hat{\mathcal{H}}_1^{(R)(\text{rf})}$ under the resonance condition $\omega^{(\text{rf})} = h_i$ can be calculated as

$$\hat{\mathcal{H}}_1^{(R)(\text{rf})} = -\Gamma_1 \left[\begin{pmatrix} 0 & 0 & e^{-i\phi_1} & 0 \\ 0 & 0 & 0 & e^{-i\phi_1} \\ e^{i\phi_1} & 0 & 0 & 0 \\ 0 & e^{i\phi_1} & 0 & 0 \end{pmatrix} + \gamma' \begin{pmatrix} 0 & a_{--} & 0 & 0 \\ a_{++} & 0 & 0 & 0 \\ 0 & 0 & 0 & a_{--} \\ 0 & 0 & a_{++} & 0 \end{pmatrix} \right],$$

where $a_{--} := e^{-i(h_2-h_1)t+\phi_1} + e^{-i(h_1+h_2)t-\phi_1}$ and $a_{++} := e^{i(h_2-h_1)t+\phi_1} + e^{i(h_1+h_2)t-\phi_1}$. The second term of $\hat{\mathcal{H}}_1^{(R)(\text{rf})}$ vanishes when $|h_1 + h_2|\tau, |h_2 - h_1|\tau \gg 1$. Then under these conditions, the Hamiltonian becomes

$$\hat{\mathcal{H}}_1^{(R)(\text{rf})} = -\Gamma_1 (\cos\phi_1 \hat{\sigma}_1^x + \sin\phi_1 \hat{\sigma}_1^y). \tag{19}$$

In the same way, the Hamiltonian $\hat{\mathcal{H}}_2^{(R)(rf)}$ can be calculated as

$$\hat{\mathcal{H}}_2^{(R)(rf)} = -\Gamma_2(\cos\phi_2\hat{\sigma}_2^x + \sin\phi_2\hat{\sigma}_2^y). \tag{20}$$

By taking the rotation operators on the individual sites, we can rewrite the Hamiltonians $\hat{\mathcal{H}}_1^{(R)(rf)}$ and $\hat{\mathcal{H}}_2^{(R)(rf)}$ by only the x-component of the Pauli matrix:

$$e^{i\phi_1\hat{\sigma}_1^z}\hat{\mathcal{H}}_1^{(R)(rf)}e^{-i\phi_1\hat{\sigma}_1^z} = -\Gamma_1\hat{\sigma}_1^x, \tag{21}$$

$$e^{i\phi_2\hat{\sigma}_2^z}\hat{\mathcal{H}}_2^{(R)(rf)}e^{-i\phi_2\hat{\sigma}_2^z} = -\Gamma_2\hat{\sigma}_2^x. \tag{22}$$

Then, the total Hamiltonian can be represented by the Ising model with site-dependent transverse fields:

$$\hat{\mathcal{H}}^{(R)} = -J\hat{\sigma}_1^z\hat{\sigma}_2^z - \Gamma_1\hat{\sigma}_1^x - \Gamma_2\hat{\sigma}_2^x. \tag{23}$$

It should be noted that the above procedure is not restricted for two spin system. Then, the NMR technique can be create the Ising model with site-dependent transverse fields in general.

3. Implementation Methods of Quantum Annealing

As stated in Sec. 1, the quantum annealing method is expected to be a powerful tool to obtain the best solution of optimization problems in a generic way. The quantum annealing methods can be categorized according to how to treat time-development. One is a stochastic method such as the Monte Carlo method which will be shown in Sec. 3.1. Other is a deterministic method such as mean-field type method and real-time dynamics. We will explain the mean-field type method and the method based on real-time dynamics in Secs. 3.2 and 3.3. Although in the Monte Carlo method and the mean-field type method, we introduce time-development in an artificial way, the merit of these methods is to be able to treat large-scale systems. The methods based on the Schrödinger equation can follow up real-time dynamics which occurs in real experimental systems. However, these methods can be used for very small systems and/or limited lattice geometries because of limited computer resources and characters of algorithms. Each method has strengths and limitations based on its individuality. Then when we use the quantum annealing, we have to choose implementation methods according to what we want to know. In this section, we explain three types of theoretical methods for the quantum annealing and some experimental results which relate to the quantum annealing.

3.1. *Monte Carlo Method*

In this section we review the Monte Carlo method as an implementation method of the quantum annealing. In physics, the Monte Carlo method is widely adopted for analysis of equilibrium properties of strongly correlated systems such as spin systems, electric systems, and bosonic systems. Originally the Monte Carlo method is used in order to calculate integrated value of given function. The simplest example is "calculation of π". Suppose we consider a square in which $-1 \leq x, y \leq 1$ and a circle whose radius is unity and center is $(x, y) = (0, 0)$. We generate pair of uniform random numbers $(-1 \leq x_i, y_i \leq 1)$ many times and calculate the following quantity:

$$\frac{\text{number of steps when } \sqrt{x_i^2 + y_i^2} \leq 1 \text{ is satisfied}}{\text{number of steps}}. \tag{24}$$

Hereafter we refer to the denominator as Monte Carlo step. The quantity should converge to $\pi/4$ in the limit of infinite Monte Carlo step. This is a pedagogical example of the Monte Carlo method. We first explain how to implement and theoretical background of the Monte Carlo method which is used in physics.

In equilibrium statistical physics, we would like to know the equilibrium value at given temperature T. The equilibrium value of the physical quantity which is represented by the operator O is defined as

$$\langle O \rangle_T^{(\text{eq})} := \frac{\text{Tr}\, O e^{-\beta \mathcal{H}}}{\text{Tr}\, e^{-\beta \mathcal{H}}}, \tag{25}$$

where Tr means the trace of matrix and β denotes the inverse temperature $\beta = (k_B T)^{-1}$. Hereafter we set the Boltzmann constant k_B to be unity. For small systems, we can obtain the equilibrium value by taking sum analytically, on the contrary, it is difficult to obtain the equilibrium value for large systems except few solvable models. Then in order to evaluate equilibrium value of the physical quantity, we often use the Monte Carlo method.

We consider the Ising model given by

$$\mathcal{H}_{\text{Ising}} = -\sum_{\langle i,j \rangle} J_{ij} \sigma_i^z \sigma_j^z - \sum_{i=1}^{N} h_i \sigma_i^z, \qquad \sigma_i^z = \pm 1. \tag{26}$$

The Ising model without transverse field can be expressed as a diagonal matrix by using "trivial" bit representation $|\uparrow\rangle$ and $|\downarrow\rangle$ which were introduced in Sec. 2. Then, in this case, we can easily calculate the eigenenergy once the eigenstate is specified.

We can use the Monte Carlo method for obtaining the equilibrium value defined by Eq. (25) as well as the calculation of π:

$$\frac{\sum_\Sigma O(\Sigma)e^{-\beta E(\Sigma)}}{\sum_\Sigma e^{-\beta E(\Sigma)}} \to \langle O \rangle_T^{(\text{eq})}, \tag{27}$$

where $O(\Sigma)$ and $E(\Sigma)$ denote the physical value of O and the eigenenergy of the eigenstate Σ, respectively. Here the eigenstate Σ is generated by uniform random number and $\sum_\Sigma 1$ is equal to Monte Carlo step. In the limit of infinite Monte Carlo step, LHS of Eq. (27) should be converge to the equilibrium value. Equilibrium statistical physics says that the probability distribution at equilibrium state can be described by the Boltzmann distribution which is proportional to $e^{-\beta E(\Sigma)}$. In this case, since we know the form of the probability distribution, it is better to use the distribution function to generate a state according to the Boltzmann distribution instead of uniform random number. This scheme is called importance sampling. When we use the importance sampling, we can obtain the equilibrium value as follows:

$$\frac{\sum_\Sigma O(\Sigma)}{\sum_\Sigma 1} \to \langle O \rangle_T^{(\text{eq})}. \tag{28}$$

In order to generate a state according to the Boltzmann distribution, we use the Markov chain Monte Carlo method. Let $P(\Sigma_a, t)$ be the probability of the a-th state at time t. In this method, time-evolution of probability distribution is given by the master equation:

$$P(\Sigma_a, t + \Delta t) = \left[\sum_{b \neq a} P(\Sigma_b, t)w(\Sigma_a|\Sigma_b) + P(\Sigma_a, t)w(\Sigma_a|\Sigma_a) \right.$$

$$\left. - \sum_{b \neq a} P(\Sigma_a, t)w(\Sigma_b|\Sigma_a) \right] \Delta t, \tag{29}$$

where $w(\Sigma_a|\Sigma_b)$ represents the transition probability from the b-th state to the a-th state in unit time. The transition probability $w(\Sigma_a|\Sigma_b)$ obeys

$$\sum_{\Sigma_a} w(\Sigma_a|\Sigma_b) = 1 \qquad (\forall \Sigma_b). \tag{30}$$

For convenience, let $\mathbf{P}(t)$ be a vector-representation of probability distribution $\{P(\Sigma_a, t)\}$. Then the master equation can be represented by

$$\mathbf{P}(t + \Delta t) = \mathcal{L}\mathbf{P}(t), \tag{31}$$

where $\mathcal{L}$ is the transition matrix whose elements are defined as

$$\mathcal{L}_{ba} := w(\Sigma_b|\Sigma_a)\Delta t, \tag{32}$$

$$\mathcal{L}_{aa} := 1 - \sum_{b\neq a}\mathcal{L}_{ba} = 1 - \sum_{b\neq a}w(\Sigma_b|\Sigma_a)\Delta t. \tag{33}$$

Here the matrix $\mathcal{L}$ is a non-negative matrix and does not depend on time. Then this time-evolution is the Markovian.

If the transition matrix $\mathcal{L}$ is prepared appropriately, which satisfies the detailed balance condition and the ergordicity, we can obtain the equilibrium probability distribution in the limit of infinite Monte Carlo step regardless of choice of the initial state because of the Perron-Frobenius theorem.

We can perform the Monte Carlo method easily as following process.

Step 1 We prepare a initial state arbitrary.

Step 2 We choose a spin randomly.

Step 3 We calculate the molecular field at the chosen site in Step 2. The molecular field at the chosen site i is defined as

$$h_i^{(\text{eff})} := \sum_j{}' J_{ij}\sigma_j^z + h_i, \tag{34}$$

where the summation takes over the nearest neighbor sites of the i-th site.

Step 4 We flip the chosen spin in Step 2 according to a probability defined by some way.

Step 5 We continue from Step 2 to Step 4 until physical quantities such as magnetization converge.

In this Monte Carlo method, we only update the chosen single spin, and thus we refer to this method as single-spin-flip method. There is an ambiguity how to define $w(\Sigma_a|\Sigma_b)$ in Step 4. Here we explain two famous choices of $w(\Sigma_a|\Sigma_b)$ as follows. Transition probability in the heat-bath method is given by

$$w_{\text{HB}}(\sigma_i^z \to -\sigma_i^z) = \frac{e^{-\beta h_i^{(\text{eff})}\sigma_i^z}}{2\cosh(\beta h_i^{(\text{eff})})}. \tag{35}$$

Transition probability in the Metropolis method is given by

$$w_{\text{MP}}(\sigma_i^z \to -\sigma_i^z) = \begin{cases} 1 & (h_i^{(\text{eff})}\sigma_i^z < 0) \\ e^{-2\beta h_i^{(\text{eff})}\sigma_i^z} & (h_i^{(\text{eff})}\sigma_i^z \geq 0) \end{cases}. \tag{36}$$

Since both two transition probabilities satisfy the detailed balance condition, the equilibrium state can be obtained definitely in the limit of infinite Monte Carlo step[a]. It is important to select how to choice the transition probability since it is known that a couple of methods can sample states in an efficient fashion.[76–83]

So far we considered the Monte Carlo method for systems where there is no off-diagonal matrix element. To perform the Monte Carlo method, in a precise mathematical sense, we only have to know how to choice the basis or appropriate transformation so as to diagonalize the given Hamiltonian. However, it is difficult to obtain equilibrium values of physical quantities of quantum systems, since we have to calculate the exponential of the given Hamiltonian $e^{-\beta\hat{\mathcal{H}}}$ in general. If we know all eigenvalues and the corresponding eigenvectors of the given Hamiltonian, we can easily calculate $e^{-\beta\hat{\mathcal{H}}}$ by the unitary transformation which diagonalizes the Hamiltonian $\hat{\mathcal{H}}$. In contrast, if we do not know all eigenvalues and eigenvectors, we have to calculate any power of the Hamiltonian $\hat{\mathcal{H}}^m$ since the matrix exponential is given by

$$e^{\hat{A}} = \sum_{m=0}^{\infty} \frac{1}{m!} \hat{A}^m. \tag{37}$$

It is difficult to calculate the matrix exponential in general. Then we have to consider the following procedure in order to use the framework of the Monte Carlo method for quantum systems.

In many cases, the Hamiltonian of quantum systems can be represented as

$$\hat{\mathcal{H}} = \hat{\mathcal{H}}_c + \hat{\mathcal{H}}_q. \tag{38}$$

Hereafter we refer to $\hat{\mathcal{H}}_c$ and $\hat{\mathcal{H}}_q$ as classical Hamiltonian and quantum Hamiltonian, respectively. The classical Hamiltonian $\hat{\mathcal{H}}_c$ is a diagonal matrix. Here we assume that $\hat{\mathcal{H}}_q$ can be easily diagonalized[b]. This is a key of the quantum Monte Carlo method as will be shown later. Since $\hat{\mathcal{H}}_c$ and $\hat{\mathcal{H}}_q$ cannot commute in general: $[\hat{\mathcal{H}}_c, \hat{\mathcal{H}}_q] \neq 0$, then $e^{-\beta\hat{\mathcal{H}}} \neq e^{-\beta\hat{\mathcal{H}}_c}e^{-\beta\hat{\mathcal{H}}_q}$. We

---

[a]Recently, the algorithm which does not use the detailed balance condition was proposed.[76,77] It should be noted that the detailed balance condition is just a necessary condition. This novel algorithm is efficient for general spin systems.

[b]This fact does not seem to be general. However we can prepare the matrices which can be easily diagonalized by the decomposition as $\hat{\mathcal{H}}_q = \sum_\ell \hat{\mathcal{H}}_q^{(\ell)}$ in many cases.

decompose the matrix exponential by introducing large integer m,

$$\exp\left(-\frac{\beta}{m}\hat{\mathcal{H}}\right) = \exp\left[-\frac{\beta}{m}(\hat{\mathcal{H}}_\mathrm{c} + \hat{\mathcal{H}}_\mathrm{q})\right]$$

$$= \exp\left(-\frac{\beta}{m}\hat{\mathcal{H}}_\mathrm{c}\right)\exp\left(-\frac{\beta}{m}\hat{\mathcal{H}}_\mathrm{q}\right) + \mathcal{O}\left(\left(\frac{\beta}{m}\right)^2\right). \quad (39)$$

This is a concrete representation of the Trotter formula.[84] From now on, we refer to m as Trotter number. By using this relation, we can perform the Monte Carlo method for quantum systems. To illustrate it, we consider the Ising model with longitudinal and transverse magnetic fields. The considered Hamiltonian is given as

$$\hat{\mathcal{H}} = -\sum_{\langle i,j\rangle} J_{ij}\hat{\sigma}_i^z\hat{\sigma}_j^z - \sum_{i=1}^N h_i^z\hat{\sigma}_i^z - \Gamma\sum_{i=1}^N \hat{\sigma}_i^x = \hat{\mathcal{H}}_\mathrm{c} + \hat{\mathcal{H}}_\mathrm{q}, \quad (40)$$

$$\hat{\mathcal{H}}_\mathrm{c} := -\sum_{\langle i,j\rangle} J_{ij}\hat{\sigma}_i^z\hat{\sigma}_j^z - \sum_{i=1}^N h_i^z\hat{\sigma}_i^z, \qquad \hat{\mathcal{H}}_\mathrm{q} := -\Gamma\sum_{i=1}^N \hat{\sigma}_i^x, \quad (41)$$

where optimization problems often can be expressed by this classical Hamiltonian $\hat{\mathcal{H}}_\mathrm{c}$. The partition function of the Hamiltonian at temperature $T(= \beta^{-1})$ is given by

$$Z = \mathrm{Tr}\,e^{-\beta\hat{\mathcal{H}}} = \sum_\Sigma \left\langle \Sigma \left| e^{-\beta(\hat{\mathcal{H}}_\mathrm{c} + \hat{\mathcal{H}}_\mathrm{q})} \right| \Sigma \right\rangle. \quad (42)$$

Using Eq. (39) we obtain

$$Z = \lim_{m\to\infty} \sum_{\{\Sigma_k\},\{\Sigma_k'\}} \left\langle \Sigma_1 \left| e^{-\beta\hat{\mathcal{H}}_\mathrm{c}/m} \right| \Sigma_1' \right\rangle \left\langle \Sigma_1' \left| e^{-\beta\hat{\mathcal{H}}_\mathrm{q}/m} \right| \Sigma_2 \right\rangle$$

$$\times \left\langle \Sigma_2 \left| e^{-\beta\hat{\mathcal{H}}_\mathrm{c}/m} \right| \Sigma_2' \right\rangle \left\langle \Sigma_2' \left| e^{-\beta\hat{\mathcal{H}}_\mathrm{q}/m} \right| \Sigma_3 \right\rangle$$

$$\times \cdots$$

$$\times \left\langle \Sigma_m \left| e^{-\beta\hat{\mathcal{H}}_\mathrm{c}/m} \right| \Sigma_m' \right\rangle \left\langle \Sigma_m' \left| e^{-\beta\hat{\mathcal{H}}_\mathrm{q}/m} \right| \Sigma_1 \right\rangle, \quad (43)$$

where $|\Sigma_k\rangle$ represents the direct-product space of N spins:

$$|\Sigma_k\rangle := |\sigma_{1,k}^z\rangle \otimes |\sigma_{2,k}^z\rangle \otimes \cdots |\sigma_{N,k}^z\rangle, \quad (44)$$

where the first and the second subscripts of $|\sigma_{i,k}^z\rangle$ indicate coordinates of the real space and the Trotter axis, respectively. Here $|\sigma_{i,k}^z\rangle = |\uparrow\rangle$ or $|\downarrow\rangle$. Equation (42) consists of two elements $\langle\Sigma_k|e^{-\beta\hat{\mathcal{H}}_\mathrm{c}/m}|\Sigma_k'\rangle$ and

$\langle \Sigma'_k | e^{-\beta \hat{\mathcal{H}}_q/m} | \Sigma_{k+1} \rangle$. Since the classical Hamiltonian $\hat{\mathcal{H}}_c$ is a diagonal matrix, the former is easily calculated:

$$\left\langle \Sigma_k \, \left| \, e^{-\beta \hat{\mathcal{H}}_c/m} \, \right| \, \Sigma'_k \right\rangle$$

$$= \exp\left[\frac{\beta}{m} \left(\sum_{\langle i,j \rangle} J_{ij} \sigma^z_{i,k} \sigma^z_{j,k} + \sum_{i=1}^N h_i \sigma^z_{i,k} \right) \right] \prod_{i=1}^N \delta(\sigma^z_{i,k}, \sigma'^z_{i,k}), \quad (45)$$

where $\sigma^z_{i,k} = \pm 1$. On the other hand, the latter $\left\langle \Sigma'_k \, \left| \, e^{-\beta \hat{\mathcal{H}}_q/m} \, \right| \, \Sigma_{k+1} \right\rangle$ is calculated as

$$\left\langle \Sigma'_k \, \left| \, e^{-\beta \hat{\mathcal{H}}_q/m} \, \right| \, \Sigma_{k+1} \right\rangle$$

$$= \left[\frac{1}{2} \sinh\left(\frac{2\beta\Gamma}{m} \right) \right]^{N/2} \exp\left[\frac{1}{2} \ln \coth \left(\frac{\beta\Gamma}{m} \sum_{i=1}^N \sigma'^z_{i,k} \sigma^z_{i,k+1} \right) \right]. \quad (46)$$

Then the partition function given by Eq. (43) can be represented as

$$Z = \lim_{m \to \infty} A \sum_{\{\sigma^z_{i,k} = \pm 1\}} \exp \left\{ \sum_{k=1}^m \left[\sum_{\langle i,j \rangle} \left(\frac{\beta J_{ij}}{m} \sigma^z_{i,k} \sigma^z_{j,k} \right) + \sum_{i=1}^N \frac{\beta h_i}{m} \sigma^z_{i,k} \right. \right.$$

$$\left. \left. + \sum_{i=1}^N \frac{1}{2} \ln \coth \left(\frac{\beta\Gamma}{m} \right) \sigma^z_{i,k} \sigma^z_{i,k+1} \right] \right\}, \quad (47)$$

where A is just a parameter which does not affect physical quantities. It should be noted that the partition function of the d-dimensional Ising model with transverse field $\hat{\mathcal{H}}$ is equivalent to that of the $(d+1)$-dimensional Ising model *without* transverse field $\mathcal{H}_{\text{eff}}$ which is given by

$$\mathcal{H}_{\text{eff}} = -\sum_{\langle i,j \rangle} \sum_{k=1}^m \frac{J_{ij}}{m} \sigma^z_{i,k} \sigma^z_{j,k} - \sum_{i=1}^N \sum_{k=1}^m \frac{h_i}{m} \sigma^z_{i,k}$$

$$- \frac{1}{\beta} \sum_{i=1}^N \sum_{k=1}^m \frac{1}{2} \ln \coth \left(\frac{\beta\Gamma}{m} \right) \sigma^z_{i,k} \sigma^z_{i,k+1}. \quad (48)$$

The coefficient of the third term of RHS is always negative, and thus the interaction along the Trotter axis is always ferromagnetic. This ferromagnetic interaction becomes strong as the value of Γ decreases. This is called the Suzuki-Trotter decomposition.[84,85]

So far we explained the Monte Carlo method as a tool for obtaining the equilibrium state. However we can also use this method to investigate stochastic dynamics of strongly correlated systems, since the Monte Carlo method is originally based on the master equation. In terms of optimization problem, our purpose is to obtain the ground state of the given Hamiltonian. Then we decrease transverse field gradually and obtain a solution. There are many Monte Carlo studies in which the quantum annealing succeeds to obtain a better solution than that by the simulated annealing.[5,8–10,12,14,86]

3.2. Deterministic Method Based on Mean-Field Approximation

In the previous section, we considered the Monte Carlo method in which time-evolution is treated as stochastic dynamics. In this section, on the other hand, we explain a deterministic method based on mean-field approximation according to Refs. [87,88]. Before we consider the quantum annealing based on the mean-field approximation, we treat the Ising model with random interactions and site-dependent longitudinal fields given by

$$\mathcal{H}_{\text{Ising}} = -\sum_{\langle i,j \rangle} J_{ij} \sigma_i^z \sigma_j^z - \sum_{i=1}^{N} h_i \sigma_i^z. \tag{49}$$

When the transverse field is absent, the molecular field of the i-th spin is given by Eq. (34). Then an equation which determines expectation value of the i-th spin at temperature $T(= \beta^{-1})$ is given by

$$m_i^z = \frac{e^{\beta h_i^{(\text{eff})}} - e^{-\beta h_i^{(\text{eff})}}}{e^{\beta h_i^{(\text{eff})}} + e^{-\beta h_i^{(\text{eff})}}} = \tanh(\beta h_i^{(\text{eff})}). \tag{50}$$

In the mean-field level, we approximate that the state σ_j^z is equal to the expectation value m_j^z in Eq. (34), and we obtain

$$m_i^z = \tanh\left[\beta\left(\sum_j{}' J_{ij} m_j^z + h_i\right)\right], \tag{51}$$

which is often called self-consistent equation.

We can obtain equilibrium value in the mean-field level by iterating the following equation until convergence:

$$m_i^z(t+1) = \tanh(\beta h_i^{(\text{eff})}(t)), \qquad h_i^{(\text{eff})}(t) = \sum_j{}' J_{ij} m_j^z(t) + h_i. \tag{52}$$

In order to judge the convergence, we introduce a distance which represents difference between the state at t-th step and that at $(t+1)$-th step as follows:

$$d(t) := \frac{1}{N} \sum_{i=1}^{N} |m_i^z(t+1) - m_i^z(t)|. \tag{53}$$

When the quantity $d(t)$ is less than a given small value (typically $\sim 10^{-8}$ or more smaller value), we judge that the calculation is converged. We summarize this method:

Step 1 We prepare a initial state arbitrary.
Step 2 We choose a spin randomly.
Step 3 We calculate the molecular field given by Eq. (34) at the chosen site in Step 2.
Step 4 We change the value of the chosen spin in Step 2 according to the obtained molecular field in Step 3.
Step 5 We continue from Step 2 to Step 4 until the distance $d(t)$ converges to small value.

The differences between the Monte Carlo method and this method are Step 4 and Step 5. We can perform the simulated annealing by decreasing temperature and using the state obtained in Step 5 as the initial state in Step 1 at the time changing temperature[c].

Next we explain a quantum version of this method. Here we apply transverse field as a quantum field. We consider the Hamiltonian given by

$$\hat{\mathcal{H}} = -\sum_{\langle i,j \rangle} J_{ij}\hat{\sigma}_i^z \hat{\sigma}_j^z - \sum_{i=1}^{N} h_i \hat{\sigma}_i^z - \Gamma \sum_{i=1}^{N} \hat{\sigma}_i^x. \tag{54}$$

The density matrix of the equilibrium state is

$$\hat{\rho} = \frac{\exp(-\beta\hat{\mathcal{H}})}{\text{Tr}\,\exp(-\beta\hat{\mathcal{H}})} = \frac{\sum_{n=1}^{2^N} \exp(-\beta\epsilon_n)\,|\lambda_n\rangle\langle\lambda_n|}{\sum_{n=1}^{2^N} \exp(-\beta\epsilon_n)}, \tag{55}$$

where ϵ_n and $|\lambda_n\rangle$ denote the n-th eigenenergy and the corresponding eigenvector. The density matrix satisfies the variational principle that minimizes free energy:

$$F = \min_{\hat{\rho}} \left[\text{Tr}\,(\hat{\mathcal{H}} + \beta^{-1}\ln\hat{\rho})\hat{\rho} \right], \tag{56}$$

where the logarithm of the matrix is defined by the series expansion as well as the definition of the matrix exponential (see Eq. (37)). Since it is difficult to obtain the density matrix, we have to consider alternative strategy as follows.

---

[c]If we want to decrease temperature rapidly, we choose not so small value for judgement of convergence.

A reduced density matrix is defined as

$$\hat{\rho}_i := \mathrm{Tr}'\,\hat{\rho} = \frac{1}{2}\left(\hat{I} + m_i^z\hat{\sigma}^z + m_i^x\hat{\sigma}^x\right), \tag{57}$$

where Tr' indicates trace over spin states except the i-th spin. The values m_i^z and m_i^x are calculated by

$$m_i^z = \mathrm{Tr}\,(\hat{\sigma}_i^z\hat{\rho}), \qquad m_i^x = \mathrm{Tr}\,(\hat{\sigma}_i^x\hat{\rho}). \tag{58}$$

The reduced density matrix satisfies the following relations:

$$\mathrm{Tr}\,(\hat{\rho}_i) = 1, \qquad \mathrm{Tr}\,(\hat{\sigma}_i^z\hat{\rho}_i) = m_i^z, \qquad \mathrm{Tr}\,(\hat{\sigma}_i^x\hat{\rho}_i) = m_i^x. \tag{59}$$

Here we assume that the density matrix can be represented by direct products of the reduced density matrices:

$$\hat{\rho} \simeq \prod_{i=1}^{N}\hat{\rho}_i, \tag{60}$$

which is mean-field approximation (in other words, decoupling approximation). Then, the free energy is expressed as

$$F \simeq \min_{\{\hat{\rho}_i\}} \mathcal{F}(\{\hat{\rho}_i\}), \tag{61}$$

$$\mathcal{F}(\{\hat{\rho}_i\}) = -\sum_{\langle i,j\rangle} J_{ij}m_i^z m_j^z - \sum_{i=1}^{N} h_i m_i^z - \Gamma\sum_{i=1}^{N} m_i^x$$

$$+\beta^{-1}\sum_{i=1}^{N}\mathrm{Tr}\,(\hat{\rho}_i\ln\hat{\rho}_i). \tag{62}$$

From the variation of $\mathcal{F}(\{\hat{\rho}_i\})$ under the normalization condition, we obtain the following relations:

$$\hat{\rho}_i = \frac{\exp(-\beta\hat{\mathcal{H}}_i)}{\mathrm{Tr}\,[\exp(-\beta\hat{\mathcal{H}}_i)]}, \tag{63}$$

$$\hat{\mathcal{H}}_i = \begin{pmatrix} -h_i - \sum_j' J_{ij}m_j^z & -\Gamma \\ -\Gamma & +h_i + \sum_j' J_{ij}m_j^z \end{pmatrix}. \tag{64}$$

Then the reduced density matrix is represented by using the n-th ($n = 1,2$) eigenvalues $\epsilon_n^{(i)}$ and the corresponding eigenvectors $|\lambda_n^{(i)}\rangle$ of $\hat{\mathcal{H}}_i$:

$$\hat{\rho}_i = \frac{\exp(-\beta\epsilon_1^{(i)})\,|\lambda_1^{(i)}\rangle\langle\lambda_1^{(i)}| + \exp(-\beta\epsilon_2^{(i)})\,|\lambda_2^{(i)}\rangle\langle\lambda_2^{(i)}|}{\exp(-\beta\epsilon_1^{(i)}) + \exp(-\beta\epsilon_2^{(i)})}. \tag{65}$$

We can also obtain the equilibrium values of physical quantities as well as the case for $\Gamma = 0$:

$$m_i^z(t+1) = \text{Tr}(\hat{\sigma}_i^z \hat{\rho}_i(t)), \qquad m_i^x(t+1) = \text{Tr}(\hat{\sigma}_i^x \hat{\rho}_i(t)), \qquad (66)$$

$$\hat{\rho}_i(t) = \frac{\exp(-\beta \hat{\mathcal{H}}_i(t))}{\text{Tr} \exp(-\beta \hat{\mathcal{H}}_i(t))}, \qquad (67)$$

$$\hat{\mathcal{H}}_i(t) = \begin{pmatrix} -h_i - \sum_j' J_{ij} m_j^z(t) & -\Gamma \\ -\Gamma & +h_i + \sum_j' J_{ij} m_j^z(t) \end{pmatrix}. \qquad (68)$$

We continue the above self-consistent equation until the following distance converges:

$$d(t) := \frac{1}{2N} \sum_{i=1}^N \left(|m_i^z(t+1) - m_i^z(t)| + |m_i^x(t+1) - m_i^x(t)| \right). \qquad (69)$$

If the temperature is zero, the reduced density matrix should be

$$\hat{\rho}_i = |\lambda_1^{(i)}\rangle \langle \lambda_1^{(i)}|, \qquad (70)$$

where we consider the case for $\epsilon_1^{(i)} < \epsilon_2^{(i)}$. Note that if and only if $-h_i - \sum_j' J_{ij} m_j^z = \Gamma = 0$, $\epsilon_1^{(i)} = \epsilon_2^{(i)}$ is satisfied. Then if we perform the quantum annealing at $T = 0$, we have to know only the ground state of the local Hamiltonian $\hat{\mathcal{H}}_i$. The procedure is the same as the case for finite temperature. By using the method, we can obtain a better solution than that obtained by the simulated annealing for some optimization problems. Recently, other type of implementation method based on mean-field approximation was proposed.[13] The method is a quantum version of the variational Bayes inference.[89] We can also obtain a better solution than the conventional variational Bayes inference.

3.3. Real-Time Dynamics

In Sec. 3.1 and Sec. 3.2, we considered artificial time-development rules such as the Markov chain Monte Carlo method and mean-field dynamics. In this section, we explain real-time dynamics which is expressed by the time-dependent Schrödinger equation:

$$i \frac{\partial}{\partial t} |\psi(t)\rangle = \hat{\mathcal{H}}(t) |\psi(t)\rangle, \qquad (71)$$

where $\hat{\mathcal{H}}(t)$ and $|\psi(t)\rangle$ denote the time-dependent Hamiltonian and the wave function at time t, respectively. The solution of this equation is given by

$$|\psi(t)\rangle = \exp\left[-i\int_0^t \hat{\mathcal{H}}(t')\mathrm{d}t'\right]|\psi(t=0)\rangle. \tag{72}$$

If we use the time-dependent Hamiltonian including time-dependent quantum field, we can perform the quantum annealing by decreasing the quantum field gradually. To obtain the solution, it is necessary to decide the initial state for Eq. (72). Since our purpose is to obtain the ground state of the given Hamiltonian which represents the optimization problem, we have no way to know the preferable initial state that leads to the ground state definitely in the adiabatic limit. However, in general, we often use a "trivial state" as the initial state. Actually, it goes well in many cases. For instance, when we consider the Ising model with time-dependent transverse field which is given by

$$\hat{\mathcal{H}}(t) = -\sum_{i,j} J_{ij}\hat{\sigma}_i^z\hat{\sigma}_j^z - \Gamma(t)\sum_{i=1}^N \hat{\sigma}_i^x, \tag{73}$$

we set the ground state for large Γ as the initial state, hence the initial state is set as

$$|\psi(t=0)\rangle = |\rightarrow\rightarrow\cdots\rightarrow\rangle, \tag{74}$$

where $|\rightarrow\rangle$ denotes the eigenstate of $\hat{\sigma}^x$:

$$|\rightarrow\rangle := \frac{1}{\sqrt{2}}(|\uparrow\rangle + |\downarrow\rangle). \tag{75}$$

In real-time dynamics, in order to obtain the ground state by using given initial condition, it is important whether there is level crossing. If there is no level crossing, the system can necessarily reach the ground state by the quantum annealing in the adiabatic limit. To show this fact, we first consider a single spin system under time-dependent longitudinal magnetic field. The Hamiltonian is given by

$$\hat{\mathcal{H}}_{\mathrm{single}}(t) = -h(t)\hat{\sigma}^z = \begin{pmatrix} -h(t) & 0 \\ 0 & h(t) \end{pmatrix}. \tag{76}$$

Suppose we set $|\psi(0)\rangle = |\downarrow\rangle$ as the initial state. For arbitrary sweeping schedules, the state at arbitrary positive t is obtained by

$$|\psi(t)\rangle = \exp\left[-i\int_0^t \hat{\mathcal{H}}_{\mathrm{single}}(t')\mathrm{d}t'\right]|\psi(0)\rangle = |\downarrow\rangle. \tag{77}$$

This is because the state $|\downarrow\rangle$ is the eigenstate of the instantaneous Hamiltonian for arbitrary time t. In general, when there is a good quantum number

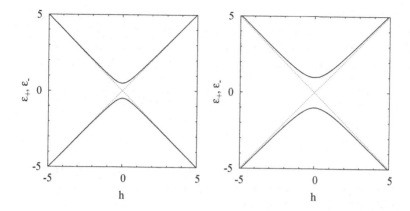

Fig. 2. Eigenenergies of the single spin system under longitudinal and transverse magnetic fields for $\Gamma = 0.5$ (left panel) and $\Gamma = 1$ (right panel). The dotted lines represent eigenenergies for $\Gamma = 0$.

and the initial state is set to be the corresponding eigenstate, the good quantum number is conserved. Then when we perform the quantum annealing method based on the real-time dynamics, we should take care of the symmetries of the considered Hamiltonian. From this, we can obtain the ground state of the considered system in the adiabatic limit if there is no level crossing. In practice, however, since we change magnetic field with finite speed, a nonadiabatic transition is inevitable. To show this fact, we consider a single spin system under longitudinal and transverse magnetic fields. The Hamiltonian of this system is given by

$$\hat{\mathcal{H}}_{\text{single}} = -h\hat{\sigma}^z - \Gamma\hat{\sigma}^x = \begin{pmatrix} -h & -\Gamma \\ -\Gamma & h \end{pmatrix}. \tag{78}$$

Since the eigenenergies are $\epsilon_{\pm} = \pm\sqrt{h^2 + \Gamma^2}$, the smallest value of the energy difference between the ground state and the excited state is 2Γ at $h = 0$ as shown in Fig. 2.

Suppose we consider the single spin system under time-dependent longitudinal magnetic field and fixed transverse magnetic field. The Hamiltonian is given by

$$\hat{\mathcal{H}}_{\text{single}}(t) = -h(t)\hat{\sigma}^z - \Gamma\hat{\sigma}^x = \begin{pmatrix} -vt & -\Gamma \\ -\Gamma & vt \end{pmatrix}, \tag{79}$$

where we adopt $h(t) = vt$ as time-dependent longitudinal field. Here we set $t = -\infty$ as the initial time. The initial state is set to be the ground state of the Hamiltonian at the initial time $|\psi(t = -\infty)\rangle = |\downarrow\rangle$. The ground state at $t = +\infty$ in the adiabatic limit is $|\psi^{(\mathrm{ad})}(t = +\infty)\rangle = |\uparrow\rangle$. Then a characteristic value which represents the nature of this dynamics is a probability of staying in the ground state at $t = +\infty$ which is defined by

$$
\begin{aligned}
P_{\mathrm{stay}} &= \left\langle \psi^{(\mathrm{ad})}(t = +\infty) \,\middle|\, \exp\left[-i \int_{-\infty}^{+\infty} \hat{\mathcal{H}}_{\mathrm{single}}(t')\mathrm{d}t' \right] \,\middle|\, \psi(t = -\infty) \right\rangle \\
&= \left\langle \uparrow \,\middle|\, \exp\left[-i \int_{-\infty}^{+\infty} \hat{\mathcal{H}}_{\mathrm{single}}(t')\mathrm{d}t' \right] \,\middle|\, \downarrow \right\rangle .
\end{aligned}
\tag{80}
$$

The probability of staying in the ground state should depend on the sweeping speed v and the characteristic energy gap and can be obtained by the Landau-Zener-Stückelberg formula:[90–92]

$$
P_{\mathrm{stay}} = 1 - \exp\left[-\frac{\pi(\Delta E)^2}{4v\Delta m} \right],
\tag{81}
$$

where ΔE and Δm represent the energy gap at the avoided level-crossing point and the difference of the magnetizations in the adiabatic limit, respectively. In this case $\Delta E = 2\Gamma$ and $\Delta m = 2$.

In many cases, typical shape of energy structure can be approximated by simple systems such as the single spin system. Then the knowledge of the simple transitions such as the Landau-Zener-Stükelberg transition and the Rosen-Zener transition[93] is useful to analyze the efficiency of the quantum annealing based on the real-time dynamics.

3.4. Experiments

Transverse field response of the Ising model has been also established in experimentally.[94–103] A dipolar-coupled disordered magnet $\mathrm{LiHo}_x \mathrm{Y}_{1-x}\mathrm{F}_4$ has easy-axis anisotropy and can be represented by the Ising model.[104,105] If we apply the longitudinal magnetic field (in other words, the magnetic field is parallel to the easy-axis), phase transition does not take place.[106,107] However, when we apply the transverse magnetic field (in other words, the magnetic field is perpendicular to the easy-axis), phase transitions occur and interesting dynamical properties shown in Ref.[6] were observed. In the phase diagram of this material, there are three phases. The ferromagnetic phase appears at intermediate temperature and low transverse magnetic field, whereas at low temperature and low transverse magnetic field, the glassy critical phase[108] appears. The paramagnetic phase exists at the other region. The glassy critical phase exhibits slow relaxation in general. It found that the characteristic relaxation time obtained by ac field susceptibility for quantum cooling in which we decrease transverse field after temperature is decreased is lower than that for temperature cooling case.[6] From this result, it has been expected that the effect of the quantum fluctuation helps us to obtain the best solution of the optimization problem.

4. Optimization Problems

Optimization problems are defined by composition elements of the considered problem and real-valued cost/gain function. They are problems to obtain the best solution such that the cost/gain function takes the minimum/maximum value. In general, the number of candidate solutions increases exponentially with the number of composition elements in optimization problems. Although we can obtain the best solution by a brute force in principle, it is difficult to obtain the best solution by such a naive method in practice. Then we have to invent an innovative method for obtaining the best solution in a practical time and limited computational resource. Optimization problems can be expressed by the Ising model in many cases. Once optimization problems are mapped onto the Ising model, we can use methods that have been considered in statistical physics and computational physics such as the quantum annealing.

In the anterior half of this section, we explain the correspondence between the Ising model and the traveling salesman problem which is one of famous optimization problems. We demonstrate the quantum annealing based on the quantum Monte Carlo simulation for this problem. In the posterior half, we explain the clustering problem as the example expressed by the Potts model which is a straightforward extension of the Ising model.

4.1. *Traveling Salesman Problem*

In this section, we consider the traveling salesman problem which is one of famous optimization problems. The setup of the traveling salesman problem is as follows:

- There are N cities.
- We move from the i-th city to the j-th city where the distance between them is $\ell_{i,j}$.
- We can pass through a city only once.
- We return the initial city after we pass through all the cities.

The traveling salesman problem is to find the minimum path under above conditions. The length of a path is given by

$$L := \sum_{a=1}^{N} \ell_{c_a, c_{a+1}}, \tag{82}$$

where c_a denotes the city where we pass through at the a-th step. In the traveling salesman problem, the length of a path is a cost function. From the fourth condition, the following relation should be satisfied:

$$c_{N+1} = c_1. \tag{83}$$

In terms of mathematics, the traveling salesman problem is to find $\{c_a\}_{a=1}^{N}$ so as to minimize the path L under the above four conditions.

If the number of cities N is small, it is easy to obtain the shortest path by a brute force. We can easily find the best solution of the traveling salesman problem for $N = 6$ shown in Fig. 3. Figure 3 (a) and (b) represent a bad solution and the best solution where the length of the path L is minimum, respectively. As the number of cities increases, the traveling salesman problem becomes seriously difficult since the number of candidate solutions is $(N-1)!/2$. Then if we want to deal with the traveling salesman problem with large N, we have to adopt smart and easy practical methods such as the simulated annealing instead of a brute force. To use the simulated annealing, we map the traveling salesman problem onto the Ising model with a couple of constraints as follows.

We consider $N \times N$ two-dimensional lattice. Let $n_{i,a}$ be the microscopic state which represents the state at the i-th city at the a-th step. The value of $n_{i,a}$ can be taken either 0 or 1. If we pass through the i-th city at the

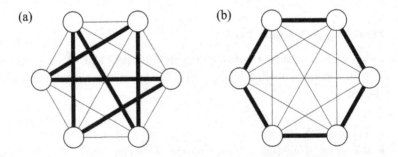

Fig. 3. Traveling salesman problem for $N = 6$. Thin lines and thick lines denote the permitted paths and selected paths, respectively. (a) Bad solution. (b) The best solution in which the length of the path is minimum.

a-th step, $n_{i,a}$ is unity whereas $n_{i,a} = 0$ if we do not pass through the i-th city at the a-th step. The third condition can be represented by

$$\sum_{a=1}^{N} n_{i,a} = 1 \quad \text{(for } \forall i\text{).} \tag{84}$$

Furthermore, since it is obvious that we can pass through only one city at the a-th step, this constraint is expressed by

$$\sum_{i=1}^{N} n_{i,a} = 1 \quad \text{(for } \forall a\text{).} \tag{85}$$

Then the length of the path L can be rewritten as

$$L = \sum_{a=1}^{N} \sum_{i,j} \ell_{i,j} n_{i,a} n_{j,a+1} = \frac{1}{4} \sum_{a=1}^{N} \sum_{i,j} \ell_{i,j} \sigma_{i,a}^{z} \sigma_{j,a+1}^{z} + \text{const.,} \tag{86}$$

where the Ising spin variable $\sigma_{i,a}^{z} = \pm 1$ is defined by

$$\sigma_{i,a}^{z} := 2 n_{i,a} - 1. \tag{87}$$

Here we used the following relation derived by Eqs. (84) and (85):

$$\sum_{a=1}^{N} \sum_{i,j} \ell_{i,j} \sigma_{i,a}^{z} = \text{const.} \tag{88}$$

Then the length of the path can be represented by the Ising spin Hamiltonian on $N \times N$ two-dimensional lattice. In general, it is difficult to obtain the stable state of the Ising model with some constraints regarded as some kind of frustration which will be shown in Sec. 5.2.

4.1.1. Monte Carlo Method

We explain how to implement the Monte Carlo method in the traveling salesman problem. We cannot use the single-spin-flip method which was explained in Sec. 3.1 because of existence of two constraints given by Eqs. (84) and (85). The simplest way of transition between states is realized by flipping four spins simultaneously as shown in Fig. 4.

Suppose we consider the case that we pass through at the i-th city at the a-th step and pass through at the j-th city at the a'-th step, which is described as

$$\sigma_{i,a}^{z} = +1, \ \sigma_{j,a}^{z} = -1, \ \sigma_{i,a'}^{z} = -1, \ \sigma_{j,a'}^{z} = +1. \tag{89}$$

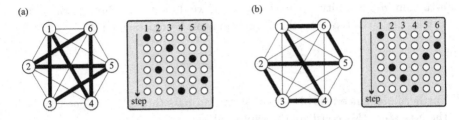

Fig. 4.   The simplest way of flipping method in traveling salesman problem. Transition between the state depicted in (a) and that depicted in (b) occurs. In this case, $i = 3$, $j = 6$, $a = 2$, and $a' = 5$.

The trial state generated by flipping four spins is as follows:

$$\sigma^z_{i,a} = -1, \ \sigma^z_{j,a} = +1, \ \sigma^z_{i,a'} = +1, \ \sigma^z_{j,a'} = -1. \tag{90}$$

The heat-bath method and the Metropolis method can be adopted for the transition probability between the present state and the trial state. In Fig. 4, $i = 3$, $j = 6$, $a = 2$, and $a' = 5$.

It should be noted that without loss of generality the initial condition can be set as

$$\sigma_{1,1} = +1, \ \sigma_{i,1} = -1 \quad (i \neq 1), \tag{91}$$

and thus we can fix the states at the first step ($a = 1$) during calculation. The number of interactions in which we try to flip all spins in each Monte Carlo step is $(N - 1)(N - 2)/2$.

4.1.2. Quantum Annealing

In order to perform the quantum annealing, we introduce the transverse field as the quantum fluctuation effect as shown in Sec. 3. The quantum Hamiltonian is given by

$$\hat{\mathcal{H}} = \frac{1}{4} \sum_{a=1}^{N} \sum_{i,j} \ell_{i,j} \hat{\sigma}^z_{i,a} \hat{\sigma}^z_{j,a+1} - \Gamma \sum_{a=1}^{N} \sum_{i=1}^{N} \hat{\sigma}^x_{i,a}, \tag{92}$$

where the first-term corresponds to the length of path and the second-term denotes the transverse field. We can map this quantum Hamiltonian on $N \times N$ two-dimensional lattice onto $N \times N \times m$ three-dimensional Ising model as well as the case which was considered in Sec. 3.1. The effective classical

Hamiltonian derived by the Suzuki-Trotter decomposition is written as

$$
\mathcal{H}_{\text{eff}} = \frac{1}{4m} \sum_{a=1}^{N} \sum_{i,j} \sum_{k=1}^{m} \ell_{i,j} \sigma_{i,a,k}^{z} \sigma_{j,a+1,k}^{z}
$$

$$
- \frac{1}{\beta} \sum_{a=1}^{N} \sum_{i=1}^{N} \sum_{k=1}^{m} \frac{1}{2} \ln \coth \left(\frac{\beta \Gamma}{m} \right) \sigma_{i,a,k}^{z} \sigma_{i,a,k+1}^{z}, \qquad \sigma_{i,a,k}^{z} = \pm 1. \quad (93)
$$

In the quantum annealing procedure, we have to take care of the constraints given by Eqs. (84) and (85) as stated before. Then the simplest way of changing state is to flip simultaneously four spins on the same layer (m is fixed) along the Trotter axis.

4.1.3. Comparison with Simulated Annealing and Quantum Annealing

In order to demonstrate the comparison with the simulated annealing and the quantum annealing, we perform the Monte Carlo simulation for the traveling salesman problem. As an example, we consider $N = 20$ cities depicted in Fig. 5 (a). The positions of these cities were generated by pair of uniform random numbers ($0 \le x_i, y_i \le 1$). The time schedules of temperature $T(t)$ for the simulated annealing and transverse field $\Gamma(t)$ for the

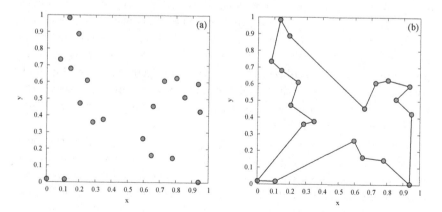

Fig. 5.   Traveling salesman problem for $N = 20$. (a) Positions of cities. (b) The best solution in which the length of the path is minimum.

quantum annealing are defined as

$$T(t) := T_0 + T_1 \left(1 - \frac{t}{\tau} \right),\tag{94}$$

$$\Gamma(t) := \Gamma_0 + \Gamma_1 \left(1 - \frac{t}{\tau} \right),\tag{95}$$

where T_0 and Γ_0 are temperature and transverse field at the final time $(t = \tau)$, and $T_0 + T_1$ and $\Gamma_0 + \Gamma_1$ are temperature and transverse field at the initial time $(t = 0)$. The value of τ^{-1} indicates the annealing speed, and the annealing speed becomes slow as the value of τ increases. In our simulations, we adopt $T_0 = \Gamma_0 = 0.01$ and $T_1 = \Gamma_1 = 5$. Furthermore, we fix the transverse field as $\Gamma = 0$ during the simulation in the simulated annealing and the temperature as $T = 0.01$ during the simulation in the quantum annealing.

We execute 100 independent simulations of simulated annealing based on the heat-bath type Monte Carlo method where each initial state generated by the uniform random number is different. To compare the efficiency of the simulated annealing and quantum annealing in an equitable manner, in the quantum annealing, the Trotter number is putted as $m = 10$, and we execute 10 independent simulations. We also calculate the minimum length of path $L_{\min}(t) := \min\{L(t')|0 \le t' \le t\}$. It should be noted that $L_{\min}(t)$ is a monotonic decreasing function. The upper panel of Fig. 6 shows the time dependence of minimum length of path $L_{\min}(t)$ for various τ. From the upper panel of Fig. 6, we can see that the convergence of minimum length of path in the quantum annealing is faster than that in the simulated annealing. We also show the sweeping time τ dependence of the minimum length of path at the final state $L_{\min}(\tau)$ in the lower panel of Fig. 6. This figure indicates that the obtained solution in the quantum annealing is always better than that in the simulated annealing. Figure 5 (b) shows the obtained best solution in both the simulated annealing and the quantum annealing with slow schedule.

In this way, we can obtain a better solution (in this case, the best solution) by both annealing methods with slow schedule. Moreover, in our calculation, the convergence of solution in the quantum annealing is faster than that in the simulated annealing, and the obtained solution in the quantum annealing is better than that in the simulated annealing regardless of sweeping time τ. Thus, we can say that the quantum annealing method is appropriate as the annealing method for the traveling salesman problem

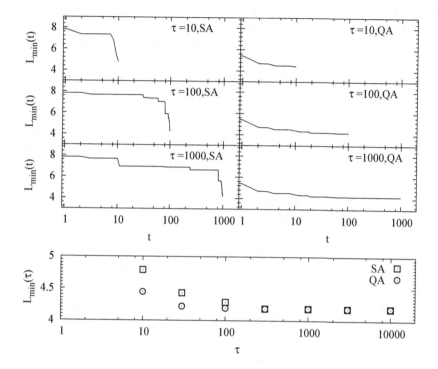

Fig. 6. (Upper panel) Time dependence of minimum length of path $L_{min}(t)$ for $\tau = 10$, 100, and 1000 obtained by the simulated annealing (SA) and the quantum annealing (QA). (Lower panel) Sweeping-time τ dependence of minimum length of path at the final state $L_{min}(\tau)$ obtained by the simulated annealing indicated by squares and the quantum annealing indicated by circles.

in comparison with the simulated annealing. This fact has been confirmed in some researches.[86,109]

4.2. Clustering Problem

In Sec. 4.1, we explained the traveling salesman problem which can be mapped onto the Ising model with some constraints. Many optimization problems can also be mapped onto the Ising model. However, there are a number of optimization problems that can be described by the other models which are straightforward extensions of the Ising model. In this section, we review the concept of clustering problem as such an example.

Clustering problem is also one of important optimization problems in information science and engineering.[12–14] We need to categorize much data in the real world according to its contents in various situations. For instance, suppose we play stock market. In order to see the socioeconomic situation, we want to extract efficiently important information related to stock market from an enormous quantity of information in news sites and newspapers. In this case, it is better to categorize many articles in news sites and newspapers according to their contents. This is an example of clustering problem which is adopted for many applications in wide area of science such as cognitive science, social science, and psychology. The clustering problem is to divide the whole set into a couple of subsets. Here we refer to the subsets as "cluster".

Figure 7 shows schematic picture of the clustering problem. Suppose we consider much data in the whole set which represents the square frame in Fig. 7 (a). The points in Fig. 7 denote individual data. In the clustering problem, our target is to find which the best division is. Figure 7 (b), (c), and (d) represent typical clustering states Σ_1, Σ_2, and Σ_*, respectively. The states Σ_1 and Σ_2 are an unstable solution and a metastable solution, respectively. The state Σ_* denotes the best solution of clustering problem.

In order to consider how to implement the quantum annealing, the clustering problem can be described by the Potts model with random interac-

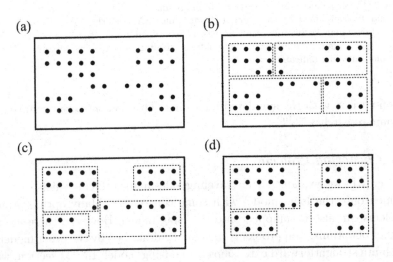

Fig. 7. Schematic pictures of clustering problem. The points represent data and the square denote the whole set. (a) Data set. (b) Unstable solution Σ_1. (c) Metastable solution Σ_2. (d) The best solution Σ_*.

tions[d]. The Hamiltonian of the Potts model is given by

$$\mathcal{H}_{\text{Potts}} = -\sum_{i,j} J_{ij} \delta_{\sigma_i, \sigma_j}, \qquad \sigma_i = 1, \cdots, Q, \tag{96}$$

where the summation runs over all pairs of the i-th and j-th data. The spin variable σ_i represents individual data. Here the value of Q represents the number of clusters. When $\sigma_i = \sigma_j$, the i-th and j-th data are in the same cluster. It is natural to adopt ferromagnetic/antiferromagnetic interaction between data in the same/different cluster. It should be noted that the Potts model is a straightforward extension of the Ising model since the Potts model is equivalent to the Ising model if $Q = 2$. Then the clustering problem is a problem to obtain the ground state of the Hamiltonian of the Potts model with given random interactions. Here we assume that the number of clusters is fixed.

Next we explain how to introduce quantum field in order to perform the quantum annealing. In optimization problems which can be represented by the Ising model, we can use transverse field as the quantum fluctuation which is represented as $-\Gamma \sum_i \sigma_i^x$. However, we cannot use this transverse field $-\Gamma \sum_i \sigma_i^x$ for the clustering problem directly, since the matrix which represents the state is $Q \times Q$ matrix. Thus, we generalize the x-component of the Pauli matrix of the Ising model as follows:

$$\hat{\tau}^x := \mathbb{E}_Q - \mathbb{I}_Q = \begin{pmatrix} 0 & -1 & -1 & \cdots & -1 \\ -1 & 0 & -1 & \cdots & -1 \\ -1 & -1 & 0 & \vdots & -1 \\ \vdots & \vdots & \vdots & \ddots & \vdots \\ -1 & -1 & -1 & \cdots & 0 \end{pmatrix}, \tag{97}$$

where $\mathbb{E}_Q$ and $\mathbb{I}_Q$ represent the $Q \times Q$ unit matrix and the $Q \times Q$ matrix whose all elements are unity. By using this generalized Pauli matrix, we can apply the quantum annealing for clustering problem.[12-14] Here we consider the following Hamiltonian:

$$\hat{\mathcal{H}} = \hat{\mathcal{H}}_{\text{Potts}} + \hat{\mathcal{H}}_{\text{q}}^{(\text{Potts})}, \qquad \hat{\mathcal{H}}_{\text{q}}^{(\text{Potts})} := -\Gamma \sum_{i=1}^{N} \hat{\tau}_i^x, \tag{98}$$

---

[d]In practice, we do not know $\{J_{ij}\}$ and have to estimate interactions when we consider the clustering problem. However, we assume the Hamiltonian for simple explanation. As shown in this section, the implementation method does not depend on the specific form of interactions.

where N is the number of individual data. As well as the case for the Ising model, we can calculate the partition function of the Hamiltonian:

$$Z_{\text{Potts}} = \text{Tr}\, e^{-\beta \hat{\mathcal{H}}} = \sum_{\Sigma} \left\langle \Sigma \left| e^{-\beta(\hat{\mathcal{H}}_{\text{Potts}} + \hat{\mathcal{H}}_q^{(\text{Potts})})} \right| \Sigma \right\rangle$$

$$= \lim_{m \to \infty} \sum_{\{\Sigma_k\},\{\Sigma'_k\}} \left\langle \Sigma_1 \left| e^{-\beta \hat{\mathcal{H}}_{\text{Potts}}/m} \right| \Sigma'_1 \right\rangle \left\langle \Sigma'_1 \left| e^{-\beta \hat{\mathcal{H}}_q^{(\text{Potts})}/m} \right| \Sigma_2 \right\rangle$$

$$\times \left\langle \Sigma_2 \left| e^{-\beta \hat{\mathcal{H}}_{\text{Potts}}/m} \right| \Sigma'_2 \right\rangle \left\langle \Sigma'_2 \left| e^{-\beta \hat{\mathcal{H}}_q^{(\text{Potts})}/m} \right| \Sigma_3 \right\rangle$$

$$\times \left\langle \Sigma_m \left| e^{-\beta \hat{\mathcal{H}}_{\text{Potts}}/m} \right| \Sigma'_m \right\rangle \left\langle \Sigma'_m \left| e^{-\beta \hat{\mathcal{H}}_q^{(\text{Potts})}/m} \right| \Sigma_1 \right\rangle, \tag{99}$$

where $|\Sigma_k\rangle$ represents the direct-product space of N spins:

$$|\Sigma_k\rangle = |\sigma_{1,k}\rangle \otimes |\sigma_{2,k}\rangle \otimes \cdots |\sigma_{N,k}\rangle. \tag{100}$$

There are two elements $\langle \Sigma_k | e^{-\beta \hat{\mathcal{H}}_{\text{Potts}}/m} |\Sigma'_k\rangle$ and $\langle \Sigma'_k | e^{-\beta \hat{\mathcal{H}}_q^{(\text{Potts})}/m} |\Sigma_{k+1}\rangle$. These factors are calculated as follows:

$$\left\langle \Sigma_k \left| e^{-\beta \hat{\mathcal{H}}_{\text{Potts}}/m} \right| \Sigma'_k \right\rangle = \exp \left(\frac{\beta}{m} \sum_{i,j} J_{ij} \delta_{\sigma_{i,k},\sigma_{j,k}} \right) \prod_{i=1}^{N} \delta_{\sigma_{i,k},\sigma'_{i,k}}, \tag{101}$$

$$\left\langle \Sigma'_k \left| e^{-\beta \hat{\mathcal{H}}_q^{(\text{Potts})}/m} \right| \Sigma_{k+1} \right\rangle = \prod_{i=1}^{N} \left[e^{-\frac{\beta \Gamma}{m} \delta_{\sigma'_{i,k}\sigma_{i,k+1}}} + \frac{1}{Q} \left(e^{-\frac{\beta \Gamma}{m}(1-Q)} - 1 \right) \right]. \tag{102}$$

By using the above expressions, we can perform the quantum Monte Carlo simulation as well as the Ising model with transverse field. If the spin variable is not $S = 1/2$ Ising spin as in the case just described, we can implement the quantum annealing by considering appropriate quantum field. There are some studies that the quantum annealing succeeds to obtain the better solution than the simulated annealing for clustering problems.[12-14]

5. Relationship between Quantum Annealing and Statistical Physics

In the preceding sections we explained the Ising model, a couple of implementation methods of the quantum annealing, and the optimization problems. There are a couple of studies that clarify the efficiency and feature of the quantum annealing in terms of statistical physics. In this section we take two examples which display relationship between quantum annealing and statistical physics focusing on the thermal fluctuation effect and the quantum fluctuation effect for ordering phenomena. In the first half, we review the Kibble-Zurek mechanism which characterizes the efficiency of the quantum annealing for systems where a second-order phase transition occurs comparing with the efficiency of the simulated annealing. In the last half, we show similarities and differences between thermal fluctuation and quantum fluctuation for frustrated Ising spin systems.

5.1. Kibble-Zurek Mechanism

In statistical physics, it has been an important topic to investigate the ordering process in systems where a phase transition takes place.[110–116] Especially, dynamical properties during changing control variables such as temperature and external fields are interesting.[111,113,115] Recently, the Kibble-Zurek mechanism has been drawing attention not only in statistical physics and condensed matter physics but also for the quantum annealing. In this section, we explain the Kibble-Zurek mechanism relating to a dynamics which passes across a second-order phase transition point. The Kibble-Zurek mechanism can make clear what happens in systems where the second-order phase transition occurs during the simulated annealing and the quantum annealing from a viewpoint of statistical physics. Before we consider the efficiency of the quantum annealing comparing with the simulated annealing by using the Kibble-Zurek mechanism, we show the general feature of the Kibble-Zurek mechanism.

As an example, we consider the Kibble-Zurek mechanism in the ferromagnetic system where the second-order phase transition occurs at finite temperature. At the second-order phase transition point, the correlation length diverges in the equilibrium state, and thus the relaxation time should be infinite. Hence, the system cannot reach the equilibrium state, when we decrease temperature to the transition temperature with finite speed. Furthermore, since the relaxation time is long around the transition temperature, it is difficult to equilibrate the system. Here, we assume that growth of correlation length stops at the temperature where the system is less able to reach the equilibrium state. If we decrease temperature slow enough, the system can reach the equilibrium state even near the transition point. Thus, it is expected that the value of stopped correlation length because of the long relaxation time depends on the annealing speed. As we will see below, the value of stopped correlation length can be scaled by the annealing speed.

To consider the second-order phase transition at finite temperature in the ferromagnetic systems, we define the dimensionless temperature g as

$$g := \frac{T - T_c}{T_c}, \tag{103}$$

where T_c is the phase transition temperature. When the absolute value of g is small, it is believed that the scaling ansatz is valid. By the scaling ansatz, the temperature-dependent correlation length $\xi(g)$ is given as[117]

$$\xi(g) \propto |g|^{-\nu}, \tag{104}$$

where ν is one of the critical exponents. Moreover, the relaxation time τ_{rel} is scaled by the following relation:[117]

$$\tau_{rel}(g) \propto [\xi(g)]^z \propto |g|^{-z\nu}, \tag{105}$$

where z is the dynamical critical exponent. Here, we decrease the temperature $T(t)$ against the time t as following schedule:

$$T(t) = T_c \left(1 - \frac{t}{\tau_Q} \right) \qquad (-\infty < t \leq \tau_Q). \tag{106}$$

The value of τ_Q^{-1} corresponds to the annealing speed. When the value of τ_Q is large/small, the system is annealed to low temperature slowly/quickly. At $t = 0$, the temperature is the phase transition temperature ($T(0) = T_c$), and the temperature is zero ($T(\tau_Q) = 0$) at $t = \tau_Q$. From Eq. (106), the dimensionless temperature g becomes the time-dependent function as follows:

$$g(t) = \frac{T(t) - T_c}{T_c} = -\frac{t}{\tau_Q}. \tag{107}$$

In the Kibble-Zurek mechanism, we assume the following situation:

$$\begin{cases} \tau_{rel}(g(t)) < |t| & : \text{system } can \text{ reach equilibrium state} \\ \tau_{rel}(g(t)) > |t| & : \text{system } cannot \text{ reach equilibrium state} \end{cases}, \tag{108}$$

where $|t|$ is a remaining time to transition temperature. That is, when a remaining time $|t|$ is longer/shorter than the relaxation time $\tau_{rel}(g(t))$, the system can/cannot reach the equilibrium state. Note that the value of considered t should be negative since the relaxation time diverges before the temperature reaches the transition temperature ($t = 0$). From this assumption, the time $\tilde{t}$ at which the system is less able to reach the equilibrium state is defined by following relation:

$$\tau_{rel}(g(\tilde{t})) = |\tilde{t}|. \tag{109}$$

Furthermore, since we have assumed that the growth of correlation length stops at $t = \tilde{t}$, the value of correlation length is always $\xi(g(\tilde{t}))$ below $T(\tilde{t})$ as shown in Fig. 8. Moreover, the dimensionless temperature at $\tilde{t}$ is expressed as

$$g(\tilde{t}) = \frac{|\tilde{t}|}{\tau_Q} = \frac{\tau_{\text{rel}}(g(\tilde{t}))}{\tau_Q} \propto \frac{|g(\tilde{t})|^{-z\nu}}{\tau_Q}. \tag{110}$$

From this relation, $g(\tilde{t})$ is scaled by the annealing speed, and from Eqs. (104) and (110), the correlation length at $t = \tilde{t}$ is scaled as follows:

$$g(\tilde{t}) \propto \tau_Q^{-\frac{1}{1+z\nu}}, \qquad \xi(g(\tilde{t})) \propto \tau_Q^{\frac{\nu}{1+z\nu}}. \tag{111}$$

Furthermore, the density of domain wall $n(t)$ is written as

$$n(t) \propto \xi(g(t))^{-d}, \tag{112}$$

where d is the spatial dimension, and $n(\tilde{t})$ at $t = \tilde{t}$ is scaled as follows:

$$n(\tilde{t}) \propto \tau_Q^{-\frac{d\nu}{1+z\nu}}. \tag{113}$$

For instance, in the ferromagnetic Ising model on two-dimensional lattice ($d = 2$, $\nu = 1$) when we adopt the Monte Carlo dynamics based on the single-spin-flip method ($z = 2.132$),[118] the correlation length and the density of domain wall at $t = \tilde{t}$ are naively obtained as

$$\xi(g(\tilde{t})) \propto \tau_Q^{0.319}, \qquad n(\tilde{t}) \propto \tau_Q^{-0.639}. \tag{114}$$

In this way, in the dynamics which passes across the second-order phase transition point at finite temperature, the correlation length and the density of domain wall (topological defect) are scaled by the annealing speed.

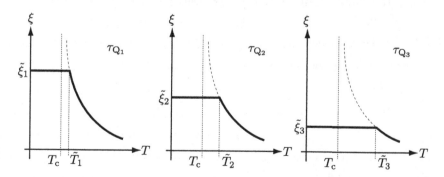

Fig. 8. Schematic of the annealing speed dependence of correlation length $\xi(g(t))$. τ_Q^{-1} is annealing speed and $\tau_{Q_1} > \tau_{Q_2} > \tau_{Q_3}$. We define $\tilde{T}_i := T_c(1 + |\tilde{t}|/\tau_{Q_i})$ and $\tilde{\xi}_i := \xi(|\tilde{t}|/\tau_{Q_i})$. The dotted curve represents correlation length in the equilibrium state.

This argument is called the Kibble-Zurek mechanism. Since the Kibble-Zurek mechanism explains the creation of topological defects induced by cooling of the system which takes place the second-order phase transition, this relates to the evolution of cosmic strings by spontaneous symmetry breaking in the Big Bang theory.[119-121] The Kibble-Zurek mechanism can also describe the creation of topological defects in magnetic models,[122,123] superfluid helium systems,[124,125] and Bose-Einstein condensations.[126,127] Next we consider the efficiency of the simulated annealing and the quantum annealing using the Kibble-Zurek mechanism by taking examples which can be treated analytically.

5.1.1. Efficiency of Simulated Annealing and Quantum Annealing

Next, we consider the efficiency of the simulated annealing and the quantum annealing according to the Kibble-Zurek mechanism. As an example, we treat the case where the non-domain wall state is the best solution. In this case, the value of $n(\tilde{t})$ approximately represents the difference between the obtained solution and the best solution. Thus, by using the Kibble-Zurek mechanism, we can compare the efficiency of annealing methods from the behavior of $n(\tilde{t})$ against the annealing speed. Suppose we solve optimization problems by using annealing methods, we would like to obtain a better solution as fast as possible, in other words, as small τ_Q as possible. Then, the comparison obtained by the Kibble-Zurek mechanism is expected to become an useful information for the optimization problems.

As an example, we consider the efficiency of the simulated annealing and the quantum annealing for the random ferromagnetic Ising chain in terms of the Kibble-Zurek mechanism according to Refs. [128,129].

5.1.2. Simulated Annealing for Random Ferromagnetic Ising Chain

The model Hamiltonian of the random ferromagnetic Ising chain is given as

$$\mathcal{H} = -\sum_i J_i \sigma_i^z \sigma_{i+1}^z, \quad \sigma_i^z = \pm 1, \tag{115}$$

where J_i is the interaction between the i-th site and the $(i+1)$-th site. The value of J_i is given by the uniform distribution between $0 < J_i \leq 1$. The distribution function $P^{(u)}(J_i)$ is given by

$$P^{(u)}(J_i) := \begin{cases} 1 & \text{for } 0 < J_i \leq 1 \\ 0 & \text{otherwise} \end{cases}. \tag{116}$$

Since the interaction J_i is always positive value, the ground state spin configuration is the all-up spin state or the all-down spin state. In this model, the ferromagnetic transition occurs at zero temperature.

The correlation function between two sites where the distance is r is written as

$$[\langle \sigma_i \sigma_{i+r} \rangle]_{\text{av}} = \left(\frac{1}{\beta} \ln \cosh \beta \right)^r, \tag{117}$$

where $\langle \cdots \rangle$ and $[\cdots]_{\text{av}}$ denote the thermal average and the random average. Physical quantities should depend on the specific spatial pattern of the random interactions $\{J_i\}$. Then, these averages are defined by

$$\langle O(\{J_i\}) \rangle := \frac{\text{Tr}\, O(\{J_i\}) e^{-\beta \mathcal{H}}}{\text{Tr}\, e^{-\beta \mathcal{H}}}, \tag{118}$$

$$[O(\{J_i\})]_{\text{av}} := \int \prod_i \mathrm{d}J_i P^{(\mathrm{u})}(J_i) O(\{J_i\}), \tag{119}$$

respectively. We omit the argument ($\{J_i\}$) for simplicity. The relationship between the correlation function and the correlation length ξ is given by

$$[\langle \sigma_i \sigma_{i+r} \rangle]_{\text{av}} = e^{-r/\xi}. \tag{120}$$

Here we mainly focus on the low-temperature limit, since the correlation length grows as temperature decreases. Then the correlation length is given as

$$\xi = -\frac{1}{\ln(\beta^{-1} \ln \cosh \beta)} \simeq \frac{\beta}{\ln 2}. \tag{121}$$

Here, we adopt the Glauber dynamics[130] as the time development, and thus the relaxation time τ_{rel} can be written as

$$\tau_{\text{rel}} = \frac{1}{1 - \tanh 2\beta} \simeq \frac{1}{2} e^{4\beta} = \frac{1}{2} e^{4\xi \ln 2}. \tag{122}$$

As we can see, in this model, the correlation length ξ and the relaxation time τ_{rel} are not the power function of temperature unlike the case of the systems where the second-order phase transition occurs at finite temperature (Eqs. (104) and (105)). This is because properties are different between phase transition which exhibits at finite temperature and that occurs at zero temperature.

We decrease temperature $T(t)$ against the time t as following schedule:

$$T(t) = -\frac{t}{\tau_Q} \qquad (-\infty < t \le 0). \tag{123}$$

Here $T_c = 0$ in this system. According to the Kibble-Zurek mechanism, we define $\tilde{t}$ by following relation:

$$\tau_{\text{rel}}(T(\tilde{t})) = |\tilde{t}|, \tag{124}$$

and, we obtain

$$T(\tilde{t}) = \frac{|\tilde{t}|}{\tau_Q} = \frac{\tau_{\text{rel}}(T(\tilde{t}))}{\tau_Q}. \tag{125}$$

By using Eqs. (121) and (122), low-temperature limit of Eq. (125) is written as

$$\frac{1}{\xi(T(\tilde{t})) \ln 2} \simeq \frac{1}{2\tau_Q} e^{4\xi(T(\tilde{t})) \ln 2}, \tag{126}$$

and, we obtain

$$\xi(T(\tilde{t})) = \frac{\ln \tau_Q + \ln 2 - \ln(\xi(T(\tilde{t})) \ln 2)}{4 \ln 2} \propto \frac{\ln \tau_Q}{4 \ln 2}. \tag{127}$$

The approximation of RHS is valid in the case of $\tau_Q \gg 1$ which indicates very slow annealing speed. Thus, we can estimate the density of domain wall $n_{\text{SA}}(\tilde{t})$ at $t = \tilde{t}$ as follows:

$$n_{\text{SA}}(\tilde{t}) \propto \frac{4 \ln 2}{\ln \tau_Q}. \tag{128}$$

5.1.3. *Quantum Annealing for Random Ferromagnetic Ising Chain*

We study the Kibble-Zurek mechanism for the random ferromagnetic Ising chain with transverse field Γ. The model Hamiltonian is given as

$$\hat{\mathcal{H}} = -\sum_i J_i \hat{\sigma}_i^z \hat{\sigma}_{i+1}^z - \Gamma \sum_i \hat{\sigma}_i^x, \tag{129}$$

where the value of J_i is given by the uniform distribution between $0 < J_i \leq 1$ as well as the case of simulated annealing. In this model, the quantum phase transition from the paramagnetic phase to the ferromagnetic phase occurs at $\Gamma_c = \exp([\ln J_i]_{\text{av}})$.[131] Here, we define the dimensionless transverse field g as

$$g := \frac{\Gamma - \Gamma_c}{\Gamma_c}. \tag{130}$$

When $|g| \ll 1$, it has been known that the correlation length obtained by the renormalization group analysis[132] is scaled by the following relation:

$$\xi(g) \propto |g|^{-\nu} \quad (\nu = 2). \tag{131}$$

Moreover, a coherence time τ_{coh} is scaled by

$$\tau_{coh}(g) \propto [\xi(g)]^z \propto |g|^{-\nu z} \qquad (\nu = 2), \tag{132}$$

where the dynamical exponent z is scaled as

$$z \propto \frac{1}{|g|}, \tag{133}$$

which is also obtained by the renormalization group analysis.[132] This means that the dynamical exponent diverges at the transition point, and this behavior is a qualitative difference between the random system and the pure system ($z = 1$). From this fact, τ_{coh} cannot be expressed by the power function of g unlike the case of the second-order phase transition at finite temperature.

We decrease transverse field $\Gamma(t)$ against the time t as following schedule:

$$\Gamma(t) = \Gamma_c \left(1 - \frac{t}{\tau_Q}\right) \qquad (-\infty < t \leq \tau_Q). \tag{134}$$

According to the Kibble-Zurek mechanism, we define $\tilde{t}$ by following relation:

$$\tau_{coh}(g(\tilde{t})) = |\tilde{t}|, \tag{135}$$

and we obtain

$$g(\tilde{t}) = \frac{|\tilde{t}|}{\tau_Q} = \frac{\tau_{coh}(g(\tilde{t}))}{\tau_Q}. \tag{136}$$

By using Eqs. (131), (132), and (133), Eq. (136) is written as

$$\frac{1}{\sqrt{\xi(g(\tilde{t}))}} \propto \frac{1}{\tau_Q}|\xi(g(\tilde{t}))|^z \propto \frac{1}{\tau_Q}|\xi(g(\tilde{t}))|^{\sqrt{\xi(g(\tilde{t}))}}, \tag{137}$$

and, we obtain

$$\left(\sqrt{\xi(g(\tilde{t}))} + \frac{1}{2}\right)\ln \xi(g(\tilde{t})) \propto \ln \tau_Q. \tag{138}$$

In the limit of $\tau_Q \gg 1$, since the value of $\xi(g(\tilde{t}))$ is very large,

$$\sqrt{\xi(g(\tilde{t}))} + \frac{1}{2} \simeq \sqrt{\xi(g(\tilde{t}))}, \tag{139}$$

and we obtain[133]

$$\xi(g(\tilde{t})) \propto \left(\frac{\ln \tau_Q}{\ln \xi(g(\tilde{t}))}\right)^2. \tag{140}$$

Moreover, since the change of $\ln \xi(g(\tilde{t}))$ is gradual in comparison with that of $\xi(g(\tilde{t}))$, we neglect $\ln \xi(g(\tilde{t}))$ and obtain

$$\xi(g(\tilde{t})) \propto (\ln \tau_Q)^2 . \tag{141}$$

From this relation, we can estimate the density of domain wall $n_{QA}(\tilde{t})$ at $t = \tilde{t}$ as follows:

$$n_{QA}(\tilde{t}) \propto (\ln \tau_Q)^{-2} . \tag{142}$$

5.1.4. Comparison between Simulated and Quantum Annealing Methods

We have shown analysis of the domain wall density in the random ferromagnetic Ising chain during the simulated annealing and the quantum annealing by the Kibble-Zurek mechanism. The obtained densities of domain wall are

$$n_{SA}(\tilde{t}) \propto (\ln \tau_Q)^{-1} \quad : \quad \text{simulated annealing}, \tag{143}$$

$$n_{QA}(\tilde{t}) \propto (\ln \tau_Q)^{-2} \quad : \quad \text{quantum annealing}. \tag{144}$$

From these relations, it is clear that the decay of $n_{QA}(\tilde{t})$ is faster than that of $n_{SA}(\tilde{t})$ against the value of τ_Q. Thus, from the Kibble-Zurek mechanism, it is concluded that the quantum annealing method is appropriate as the annealing method for the random ferromagnetic Ising chain in comparison with the simulated annealing method. Suppose we consider the ferromagnetic Ising chain with homogeneous interaction ($J_i = 1$ for all i). In this case, both the domain wall density in the simulated annealing and that in the quantum annealing are obtained as

$$n(\tilde{t}) \propto \frac{1}{\sqrt{\tau_Q}}. \tag{145}$$

This relation for the simulated annealing can be obtained by a simple calculation as well as the case of the random Ising spin chain. On top of that, the relation for the quantum annealing can be derived by Eq. (113). Here the critical exponent ν of the transverse Ising chain with homogeneous interaction is $\nu = 1$ and the dynamical exponent of this system is $z = 1$. Then there is no difference between the simulated annealing and the quantum annealing in the case of the homogeneous ferromagnetic Ising chain. However, since the optimization problem has some kind of randomness, the abovementioned result encourages that the quantum annealing is better than the simulated annealing for optimization problems.

In general, the existence of the phase transition in optimization problems negatively influences performance of annealing methods. Here, we have introduced the Kibble-Zurek mechanism relating to the dynamics which passes across the second-order phase transition point. As the specific example, we have analyzed the efficiencies of the simulated annealing and the quantum annealing for the random ferromagnetic Ising chain according to the Kibble-Zurek mechanism. For this model, the efficiency of the quantum annealing is better than that of the simulated annealing. Of course, since the efficiency of annealing methods depends on the details of optimization problems, it is not to say that the quantum annealing is always appropriate as the annealing method for general optimization problems in comparison with the simulated annealing. Moreover, we have to develop a theory based on the Kibble-Zurek mechanism itself,[134] since we assume the growth of the correlation length stops at $t > \tilde{t}$. For example, if we adapt the Kibble-Zurek mechanism to two- or three-dimensional models and more complicated models, it is difficult to estimate the correlation length analytically, and thus we should execute numerical simulations such as the Monte Carlo simulation. For example, in the two-dimensional Ising model with random interactions, it has been shown that the efficiency of the quantum annealing is better than that of the simulated annealing by Monte Carlo simulation.[129] Although the efficiency of annealing methods for a number of optimization problems has been clarified by the Kibble-Zurek mechanism, it remains to be an open problem to investigate when to use the quantum annealing exhaustively.

In the above-mentioned argument, the phase transition under consideration is of the second order. What happens if we adapt the same argument for the other type phase transitions such as first-order phase transition and Kosterlitz-Thouless (KT) transition? In these phase transitions, the behaviors of correlation length are different from that in systems where a second-order phase transition occurs: the finite-correlation length at the first-order phase transition point and the quasi-long-range correlation length at the KT transition point. Thus, it is an interesting problem to clarify relationship between behaviors of correlation length and the *generalized* Kibble-Zurek mechanism. By considering dynamical nature of the optimization problems in terms of non-equilibrium statistical physics in a deeper way, we believe that the quantum annealing method will become a central part of practical method for optimization problems.

5.2. Frustration Effects for Simulated Annealing and Quantum Annealing

In many cases optimization problems can be represented by the Ising model with random interactions and magnetic fields as mentioned before. The Hamiltonian of this system is given by

$$\mathcal{H} = -\sum_{i,j} J_{ij}\sigma_i^z\sigma_j^z - \sum_{i=1}^{N} h_i\sigma_i^z, \quad \sigma_i^z = \pm 1. \tag{146}$$

When all interactions are ferromagnetic as the previous example in Sec. 5.1, the ground state is the all-up or the all-down states. However, if there are antiferromagnetic interactions in the system, the situation becomes different. In order to show the difference between ferromagnetic interaction and antiferromagnetic interaction, we first consider three spin system on triangle cluster as shown in Fig. 9. In this section, we treat the case for $h_i = 0$ for all i. The dotted and solid lines in Fig. 9 represent ferromagnetic and antiferromagnetic interactions, respectively.

The considered Hamiltonian is written as

$$\mathcal{H}_{\text{triangle}} = -J(\sigma_1^z\sigma_2^z + \sigma_2^z\sigma_3^z + \sigma_3^z\sigma_1^z). \tag{147}$$

Here we set the all interactions are the same value for simplicity. The ground states for positive J (ferromagnetic interaction) are the all-up or the all-down states shown in Fig. 9 (a). In these states, all spins between all interactions are energetically favorable states. In the case of negative J (antiferromagnetic interaction), while on the other hand, six states shown in Fig. 9 (b) are ground states. These ground states have unfavorable interactions

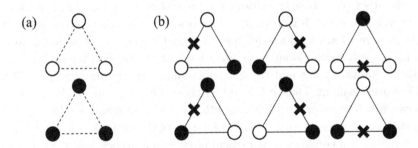

Fig. 9. Three spin system on triangle cluster. The dotted and solid lines represent ferromagnetic and antiferromagnetic interactions, respectively. The open and solid circles are the +1-state and the −1-state, respectively. The crosses indicate the positions of unfavorable interactions. (a) Ground states for ferromagnetic case. (b) Ground states for antiferromagnetic case.

indicated by the crosses in Fig. 9 (b). This situation is called frustration. In the homogeneous antiferromagnetic Ising spin systems on lattices based on triangle such as triangular lattice and kagomé lattice, frustration appears in all triangles. Since such frustration comes from lattice geometry, this is called geometrical frustration. It should be noted that the homogeneous antiferromagnetic Ising spin systems on square lattice and hexagonal lattice have no frustration. Since these systems are bipartite systems which can be decomposed by two sublattices, these systems can be transformed on the ferromagnetic systems by local gauge transformation of all spins belonging to one of the sublattices.

Frustration appears in also inhomogeneous systems as shown in Fig. 10. The squares pointed by stars in Fig. 10 represent frustration plaquettes which are satisfied following relation:

$$\kappa_k := \prod_{i,j \in \square_k} J_{ij} < 0, \tag{148}$$

where $\square_k$ indicates the smallest square plaquette at the position k. If κ_k for all k is positive, the system is not frustrated.

In general, frustration prevents the system from conventional magnetic ordering such as ferromagnetic order and Néel order, since there is no state where all interactions are satisfied energetically in frustrated systems. Frustration makes peculiar density of states which induces unconventional phase transition and slow dynamics.[112,115,135–144] Although many optimization problems can be represented by the Ising model with random interactions and magnetic fields, here we focus on the frustration effect which comes

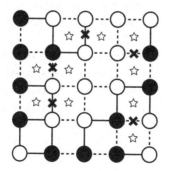

Fig. 10. A ground state of the Ising spin system with random interactions. The dotted and solid lines represent ferromagnetic and antiferromagnetic interactions, respectively. The open and solid circles are the +1-state and the −1-state, respectively. The stars and crosses indicate frustration plaquettes and unfavorable interactions, respectively.

from non-random interactions. In terms of statistical physics, this is a first-step study to investigate similarities and differences between thermal fluctuation and quantum fluctuation for frustrated systems. Furthermore, it is important topic for the optimization problems to consider the thermal fluctuation and quantum fluctuation effects for frustrated systems. To obtain the ground state of frustrated systems is to find how to put the unsatisfied bonds represented by the crosses. Since the unsatisfied bonds are regarded as some kind of constraints, this situation is similar with the traveling salesman problem in which there are some constraints as mentioned before. We explain two topics in this section. In the first half, we consider the order by disorder effect in fully-frustrated systems. In the last half, we explain non-monotonic dynamics in decorated bond systems.

5.2.1. Thermal Fluctuation and Quantum Fluctuation Effect of Geometrical Frustrated Systems

In general, there are many degenerated ground states in geometrical frustrated systems such as triangular antiferromagnetic Ising spin systems and kagomé antiferromagnetic Ising spin systems. In these cases, non-zero residual entropy which is entropy at zero temperature exists. Typical configurations of ground states of the triangular antiferromagnetic Ising spin systems are shown in Fig. 11. The residual entropy per spin of this system is $S_{res}^{(tri)} \simeq 0.323 k_B$,[145-148] where k_B is the Boltzmann constant. Since the total entropy per spin is $k_B \ln 2 \simeq 0.693 k_B$, 46.6% of the total entropy remains even at zero temperature. In other words, there are macroscopic degenerated ground states in this system. In the antiferromagnetic Ising spin system on kagomé lattice, there are also macroscopic degenerated ground states. The residual entropy per spin of this system is $S_{res}^{(kag)} \simeq 0.502 k_B$, which is 72.4% of the total entropy.[149]

Suppose we apply the simulated annealing or the quantum annealing with slow schedule for geometrical frustrated spin systems. Since there are macroscopically degenerated ground states in these systems, our purpose is to clarify whether all ground states are obtained with the same probabilities or biased probabilities. We first consider the obtained ground states in the case of the simulated annealing with slow schedule. If we decrease temperature slow enough, the obtained state should satisfy the equilibrium probability distribution. When the temperature is $k_B T \ll |J|$, the equilibrium probabilities of the ground states are dominant and that of any excited states can be neglected. The principle of equal weight which is the keystone in the equilibrium statistical physics says that if the eigenenergies of the

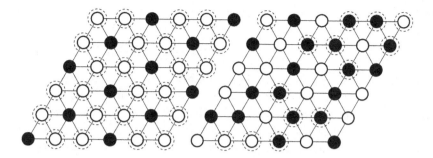

Fig. 11. Typical configurations of ground states of antiferromagnetic Ising spin system on triangular lattice. The open and solid circles are the $+1$-state and the -1-state, respectively. The dotted circles indicate free spin where the molecular field is zero.

microscopic state Σ_A and Σ_B are the same, the equilibrium probability of Σ_A and that of Σ_B are also the same. Then we obtain all macroscopic degenerated ground states with the same probability after the simulated annealing with slow schedule.

Next we consider the obtained ground states in the case of the quantum annealing where the transverse field decreases slow enough. Here we assume that the initial state is set to be the ground state of the Hamiltonian at the initial time. In order to capture the feature of the ground states in a graphical way, it is convenient to introduce the concept of free spin where the molecular field is zero. The molecular field at the i-th site is given by

$$h_i^{(\text{eff})} := J \sum_j{}' \sigma_j^z, \tag{149}$$

where the summation runs over the nearest-neighbor sites of the i-th site. For instance, in Fig. 11, spins indicated by dotted circles are free spins. Here, the transverse field is expressed as

$$-\Gamma \sum_i \hat{\sigma}_i^x = -\Gamma \sum_i (\hat{\sigma}_i^+ + \hat{\sigma}_i^-), \tag{150}$$

where $\hat{\sigma}_i^+$ and $\hat{\sigma}_i^-$ denote the raising and lowering operators at the i-th site, respectively. They are defined by

$$\hat{\sigma}^+ := \begin{pmatrix} 0 & 1 \\ 0 & 0 \end{pmatrix}, \qquad \hat{\sigma}^- := \begin{pmatrix} 0 & 0 \\ 1 & 0 \end{pmatrix}. \tag{151}$$

The x-component of the Pauli matrix corresponds to the operator which flips the considered spin:

$$\hat{\sigma}^x \left| \uparrow \right\rangle = \left| \downarrow \right\rangle, \qquad \hat{\sigma}^x \left| \downarrow \right\rangle = \left| \uparrow \right\rangle. \tag{152}$$

From this, the states which have large number of free spins are expected to become stable at the limit of $\Gamma \to 0+$ and $T = 0$. Actually, in the adiabatic limit, the amplitudes of the states which have the maximum number of free spins are larger than the others.[150–154] When we decrease the transverse field slow enough, the state at each time can be well approximated by the ground state of the instantaneous Hamiltonian. Then we obtain specific ground states with high probability after the quantum annealing with slow schedule.

In this section, we considered the thermal fluctuation effect and the quantum fluctuation effect in the adiabatic limit. The simulated annealing can obtain all the ground states with the same probability, while on the other hand, the quantum annealing can obtain specific ground states in this limit. The biased probability distribution can be explained by the character of the quantum Hamiltonian. The selected states should depend on how to choice the quantum Hamiltonian. When we adopt the exchange type interaction as the quantum field, the states that have the maximum value of the "free spin pair" should be selected. Moreover, it is an interesting topic to investigate differences between the simulated annealing and the quantum annealing with finite speed not only in terms of the quantum annealing but also in nonequilibrium statistical physics and condensed matter physics. At the present stage, to consider dynamic phenomena in strongly correlated systems is difficult, since a small number of theoretical methods for obtaining dynamic phenomena have been developed. If the technology of the artificial lattices develops more than ever, real-time dynamics and time-dependent phenomena of frustrated spin systems can be observed in real experiments.

5.2.2. Non-Monotonic Behavior of Correlation Function in Decorated Bond System

In the ferromagnetic Ising spin systems, the correlation function behaves monotonic against the temperature and transverse field. However, the behavior of the correlation function is non-monotonic as a function of temperature in some frustrated spin systems. As an example of non-monotonic correlation function, we introduce equilibrium properties of the correlation function in decorated bond systems in which the frustration exists. The

Hamiltonian of the decorated bond systems where the number of system spins is two shown in Fig. 12 is given by

$$\mathcal{H} = -J_{\text{dir}}\sigma_1^z\sigma_2^z - J\sum_{i=1}^{N_{\text{d}}} s_i^z(\sigma_1^z + \sigma_2^z), \tag{153}$$

where $\sigma_i^z = \pm 1$ and $s_i^z = \pm 1$ are, respectively, called system spins and decorated spins, and N_{d} is the number of decorated spins. The circles and the squares in Fig. 12 represent the system spins and the decorated spins, respectively.

When the direct interaction between system spins J_{dir} is zero and the decorated bond J is positive, the correlation function between system spins $\langle\sigma_1^z\sigma_2^z\rangle$ is always positive and monotonic decaying function against the temperature. When the direct interaction between system spins J_{dir} is negative and the decorated bond J is zero, on the other hand, the correlation function $\langle\sigma_1^z\sigma_2^z\rangle$ is always negative and monotonic increasing function against the temperature. From this, the correlation function $\langle\sigma_1^z\sigma_2^z\rangle$ is expected to behave non-monotonic in some cases for negative J_{dir} and positive J or positive J_{dir} and negative J. In order to obtain temperature dependence of the correlation function between system spins, we trace over spin states except the system spins:

$$\text{Tr}_{\{s_i^z\}}e^{-\beta\mathcal{H}} = Ae^{K_{\text{eff}}\sigma_1^z\sigma_2^z}, \tag{154}$$

where A is just a constant which does not affect any physical quantities and the effective coupling K_{eff} is given by

$$K_{\text{eff}} = \frac{N_{\text{d}}}{2}\ln\cosh(2\beta J) + \beta J_{\text{dir}}. \tag{155}$$

Temperature dependence of the correlation function between system spins

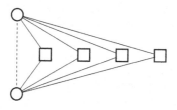

Fig. 12. Decorated bond system where the number of system spins is two and the number of decorated spins is four ($N_{\text{d}} = 4$). The circles and squares represent system spins and decorated spins, respectively. The dotted and solid lines indicate the direct interaction between system spins and the decorated bonds, respectively.

is represented by using K_{eff}:

$$C^{(c)}(T) := \langle \sigma_1^z \sigma_2^z \rangle = \frac{\text{Tr}\,\sigma_1^z \sigma_2^z e^{-\beta \mathcal{H}}}{\text{Tr}\, e^{-\beta \mathcal{H}}} = \frac{\text{Tr}\,\sigma_1^z \sigma_2^z e^{K_{\text{eff}} \sigma_1^z \sigma_2^z}}{\text{Tr}\, e^{K_{\text{eff}} \sigma_1^z \sigma_2^z}} = \tanh K_{\text{eff}}.$$

(156)

Hereafter we set J as the energy unit and J is positive. In order to compare the effect of the direct interaction J_{dir} fairly, we assume the form such as $J_{\text{dir}} = -x N_{\text{d}} J$. This is because the effective coupling K_{eff} is proportional to the number of decorated spins N_{d} under the assumption.

Figure 13 shows temperature dependence of correlation function between the system spins for $N_{\text{d}} = 1$ and $N_{\text{d}} = 10$ for several x. For small x and large x, the correlation function $C^{(c)}(T)$ is monotonic decreasing and increasing functions, respectively, against the temperature. However, the correlation function $C^{(c)}(T)$ behaves non-monotonic as a function of temperature for intermediate x. At the temperatures where the effective coupling K_{eff} is larger than the critical value of the ferromagnetic Ising spin system on square lattice[19] $K_{\text{c}}^{(\text{square})} = \frac{1}{2}\ln(1 + \sqrt{2})$, ferromagnetic phase appears. On the other hand, at the temperature where K_{eff} is less than $-K_{\text{c}}^{(\text{square})}$, antiferromagnetic phase appears. In this case, successive

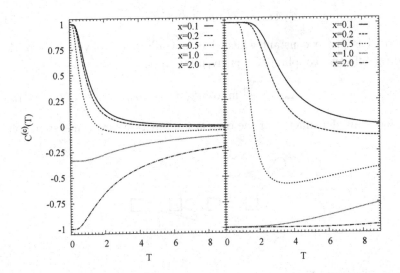

Fig. 13. The correlation function between system spins $C^{(c)}(T)$ as a function of temperature for $N_{\text{d}} = 1$ (left panel) and for $N_{\text{d}} = 10$ (right panel) in the cases of $x = 0.1$, 0.2, 0.5, 1.0, and 2.0.

phase transitions such as paramagnetic $\to$ antiferromagnetic $\to$ paramagnetic $\to$ ferromagnetic phases occur. Such phase transitions are called reentrant phase transitions which are sometimes appeared in frustrated systems.[115,139,155-160]

We consider transverse field response of the decorated bond systems in the ground state. The Hamiltonian of the decorated bond system with transverse field is expressed as

$$\hat{\mathcal{H}} = -J_{\text{dir}}\hat{\sigma}_1^z\hat{\sigma}_2^z - J\sum_{i=1}^{N_d}\hat{s}_i^z(\hat{\sigma}_1^z + \hat{\sigma}_2^z) - \Gamma(\hat{\sigma}_1^x + \hat{\sigma}_2^x + \sum_{i=1}^{N_d}\hat{s}_i^x), \quad (157)$$

where $\hat{s}_i^\alpha$ denotes the α-component of the Pauli matrix of the i-th decorated spin. Here we consider transverse-field dependence of the correlation function in the ground state given by

$$C^{(q)}(\Gamma) := \langle\psi^{(gs)}(\Gamma)|\hat{\sigma}_1^z\hat{\sigma}_2^z|\psi^{(gs)}(\Gamma)\rangle, \quad (158)$$

where $|\psi^{(gs)}(\Gamma)\rangle$ denotes the ground state at the transverse field Γ. Figure 14 shows transverse-field dependence of $C^{(q)}(\Gamma)$ for $N_d = 1$ and $N_d = 10$ for several x.

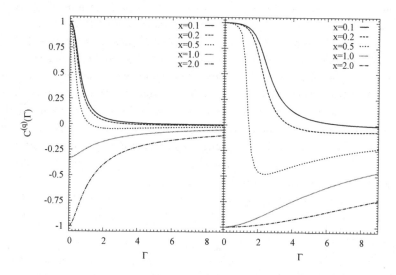

Fig. 14. The correlation function between system spins $C^{(q)}(\Gamma)$ as a function of transverse field for $N_d = 1$ (left panels) and for $N_d = 10$ (right panels) in the cases of $x = 0.1$, 0.2, 0.5, 1.0, and 2.0.

For small x and large x, the correlation function $C^{(q)}(\Gamma)$ behaves monotonic decreasing and increasing, respectively as a function of transverse field, whereas for intermediate x, transverse-field dependence of the correlation function behaves nonmonotonic as well as the case of thermal fluctuation. Then, the reentrant phase transition also occurs by changing the transverse field. However there is a difference between the thermal fluctuation effect and the quantum fluctuation effect for decorated bond system. The temperature where $C^{(c)}(T) = 0$ is satisfied is the same when we change the number of decorated spins N_d, whereas the transverse field at $C^{(q)}(\Gamma) = 0$ is different when N_d is changed.

The thermal fluctuation and the quantum fluctuation have similar properties for the phase transition phenomena in general. Indeed, the reentrant phase transitions occur by changing the thermal fluctuation and also the quantum fluctuation as shown in this section. However as described in Sec. 5.1, in order to obtain the best solution of optimization problems, it is better to erase phase transition. By dealing with thermal and quantum fluctuation effects for frustrated systems exhaustively, we can construct the best form of the adding fluctuation which erases phase transition[e].

6. Conclusion

In this paper, we described some aspects of the quantum annealing from viewpoints of statistical physics, condensed matter physics, and computational physics. Originally, the quantum annealing has been proposed as a method which can solve efficiently optimization problems in a generic way. Since many optimization problems can be mapped onto the Ising model or generalized Ising model such as the clock model and the Potts model, it has been considered that we can obtain a better solution by using methods which were developed in computational physics. For instance, we can obtain a better solution by decreasing temperature (thermal fluctuation) gradually in the simulated annealing which is one of the most famous practical methods. In the quantum annealing, we decrease an introduced quantum field (quantum fluctuation) instead of temperature (thermal fluctuation). In many studies, it was reported that a better solution can be obtained

---

[e]It is not necessary that the adding fluctuation is restricted in quantum physics. From a viewpoint of optimization problems, we can arbitrary form adding term. Furthermore, it has studied that other novel fluctuation which may be able to erase phase transition as an alternative to thermal and quantum fluctuations.[14,116,161,162] Of course, if we want to realize experimentally, it is better that the added fluctuation term should be some kind of quantum fluctuation.

by the quantum annealing efficiently in comparison with the simulated annealing as we explained in Sec. 4. Thus, the quantum annealing method is expected to be a generic and powerful solver of optimization problems as an alternative to the simulated annealing.

The quantum annealing has become a milestone of some related fields under the situation in which the quantum annealing itself has been studied exhaustively. Since we use the quantum fluctuation in the quantum annealing with ingenuity, to obtain a better solution by using the quantum annealing is a kind of quantum information processing. Thus, many implementation methods of the quantum annealing in theoretical and experimental ways have been proposed by many researchers. A number of theoretical implementation methods are proposed based on knowledge of statistical physics. As we shown in Sec. 5, question of what are differences between the simulated annealing and the quantum annealing and question of which is efficient in the given optimization problem are catalysts to investigate differences between the thermal fluctuation and the quantum fluctuation in a deeper way. On top of that, studies on the quantum annealing are expected to open the door to consider equilibrium and nonequilibrium statistical physics. Recently, preparation methods of intended Hamiltonian have been established in some experimental systems such as artificial lattices and nuclear magnetic resonance because of recent development of experimental techniques. As long as we use classical computer and our present knowledge, there are a huge number of problems where to obtain the best solution is difficult without any and every approximation in theoretical methods. However if we prepare the Hamiltonian which expresses our intended problem, we can *calculate* experimentally the stable state of the prepared Hamiltonian in near future.

The quantum annealing transcends just a method for obtaining the best solution of optimization problems and it will make a development in wide area of science. Although it seems that studies on the quantum annealing itself have been well established, we believe that the quantum annealing plays a role as a bridge with the abovementioned area of science and the quantum information.

Acknowledgement

The authors are grateful to Bernard Barbara, Bikas K. Charkrabarti, Naomichi Hatano, Masaki Hirano, Naoki Kawashima, Kenichi Kurihara, Yoshiki Matsuda, Seiji Miyashita, Hiroshi Nakagawa, Mikio Nakahara, Hidetoshi Nishimori, Masayuki Ohzeki, Hans de Raedt, Per Arne Rikvold,

Issei Sato, Sei Suzuki, Eric Vincent, and Yoshihisa Yamamoto for their valuable comments. S.T. acknowledges Keisuke Fujii, Yoshifumi Nakada, and Takahiro Sagawa for their useful discussion during the lecture. S.T. is partly supported by Grand-in-Aid for JSPS Fellows (23-7601). R.T. is partly supported financially by National Institute for Materials Science (NIMS). The computation in the present work was performed on computers at the Suprecomputer Center, Institute for Solid State Physics, University of Tokyo.

References

1. S. Kirkpatrick, C. D. Gelatt Jr., and M. P. Vecchi, *Science* **220**, 671 (1983).
2. S. Kirkpatrick, *J. Stat. Phys.* **34**, 975 (1984).
3. S. Geman and D. Geman, *IEEE Transactions on Pattern Analysis and Machine Intelligence* **6**, 721 (1984).
4. A. B. Finnila, M. A. Gomez, C. Sebenik, C. Stenson, and J. D. Doll, *Chem. Phys. Lett.* **219**, 343 (1994).
5. T. Kadowaki and H. Nishimori, *Phys. Rev. E* **58**, 5355 (1998).
6. J. Brooke, D. Bitko, T. F. Rosenbaum, and G. Aeppli, *Science* **284**, 779 (1999).
7. E. Farhi, J. Goldstone, S. Gutmann, J. Lapan, A. Lundgren, and D. Preda, *Science* **292**, 472 (2001).
8. G. E. Santoro, R. Martoňák, E. Tosatti, and R. Car, *Science* **295**, 2427 (2002).
9. A. Das and B. K. Chakrabarti, *Quantum Annealing and Related Optimization Methods* (Springer, Heidelberg, 2005).
10. A. Das and B. K. Chakrabarti, *Rev. Mod. Phys.* **80**, 1061 (2008).
11. M. Ohzeki and H. Nishimori, *J. Comp. and Theor. Nanoscience* **8**, 963 (2011).
12. K. Kurihara, S. Tanaka, and S. Miyashita, *Proceedings of the 25th Conference on Uncertainty in Artificial Intelligence* (2009).
13. I. Sato, K. Kurihara, S. Tanaka, H. Nakagawa, and S. Miyashita, *Proceedings of the 25th Conference on Uncertainty in Artificial Intelligence* (2009).
14. S. Tanaka, R. Tamura, I. Sato, and K. Kurihara, to appear in *Kinki University Quantum Computing Series: "Summer School on Diversities in Quantum Computation/Information"*.
15. C. Jarzynski, *Phys. Rev. Lett.* **78**, 2690 (1997).
16. C. Jarzynski, *Phys. Rev. E* **56**, 5018 (1997).
17. M. Ohzeki, *Phys. Rev. Lett.* **105**, 050401 (2010).
18. E. Ising, *Z. Phys.* **31**, 253 (1925).
19. L. Onsager, *Phys. Rev.* **65**, 117 (1944).
20. M. Blume, *Phys. Rev.* **141**, 517 (1966).
21. H. W. Capel, *Phys. Lett.* **23**, 327 (1966).
22. J. Tobochnik, *Phys. Rev. B* **26**, 6201 (1982).
23. M. S. S. Challa and D. P. Landau, *Phys. Rev. B* **33**, 437 (1986).
24. R. B. Potts, *Proc. Cambridge Philos. Soc.* **48**, 106 (1952).

25. F. Y. Wu, *Rev. Mod. Phys.* **54**, 235 (1982).
26. T. Ohtsuka, *J. Phys. Soc. Jpn.* **16**, 1549 (1961).
27. M. Rayl, O. E. Vilches, and J. C. Wheatley, *Phys. Rev.* **165**, 698 (1968).
28. K. Ôno, M. Shinohara, A. Ito, N. Sakai, and M. Suenaga, *Phys. Rev. Lett.* **24**, 770 (1970).
29. N. Achiwa, *J. Phys. Soc. Jpn.* **27**, 561 (1969).
30. M. Mekata and K. Adachi, *J. Phys. Soc. Jpn.* **44**, 806 (1978).
31. A. H. Cooke, D. T. Edmonds, F. R. McKim, and W. P. Wolf, *Proc. Roy. Soc. London Ser. A* **252**, 246 (1959).
32. A. H. Cooke, D. T. Edmonds, C. B. P. Finn, and W. P. Wolf, *Proc. Roy. Soc. London Ser. A* **306**, 313 (1968).
33. A. H. Cooke, D. T. Edmonds, C. B. P. Finn, and W. P. Wolf, *Proc. Roy. Soc. London Ser. A* **306**, 335 (1968).
34. K. Takeda, M. Matsuura, S. Matsukawa, Y. Ajiro, and T. Haseda, *Proc. 12th Int. Conf. Low Temp. Phys., Kyoto* 803 (1970).
35. K. Takeda, S. Matsukawa, and T. Haseda, *J. Phys. Soc. Jpn.* **30**, 1330 (1971).
36. B. N. Figgis, M. Gerloch, and R. Mason, *Acta. Crystallogr.* **17**, 506 (1964).
37. R. F. Wielinga, H. W. J. Blote, J. A. Roest, and W. J. Huiskamp, *Physica* **34**, 223 (1967).
38. K. W. Mess, E. Lagendijk, D. A. Curtis, and W. J. Huiskamp, *Physica* **34**, 126 (1967).
39. G. R. Hoy and F. de S. Barros, *Phys. Rev.* **139**, A929 (1965).
40. M. Matsuura, H. W. J. Blote, and W. J. Huiskamp, *Physica* **50**, 444 (1970).
41. R. D. Pierce and S. A. Friedberg, *Phys. Rev. B* **3**, 934 (1971).
42. K. Takeda and S. Matsukawa, *J. Phys. Soc. Jpn.* **30**, 887 (1971).
43. E. Stryjewski and N. Giordano, *Adv. Phys.* **26**, 487 (1977).
44. D. J. Breed, K. Gilijamse, and A. R. Miedema, *Physica* **45**, 205 (1969).
45. K. Ôno, A. Ito, and T. Fujita, *J. Phys. Soc. Jpn.* **19**, 2119 (1964).
46. R. J. Birgeneau, W. B. Yelon, E. Cohen, and J. Makovsky, *Phys. Rev. B* **5**, 2607 (1972).
47. J. C. Wright, H. W. Moos, J. H. Colwell, B. W. Magnum, and D. D. Thornton, *Phys. Rev. B* **3**, 843 (1971).
48. G. T. Rado, *Phys. Rev. Lett.* **23**, 644 (1969).
49. W. Scharenberg and G. Will, *Int. J. Magnetism* **1**, 277 (1971).
50. H. Fuess, A. Kallel, and F. Tchéou, *Solid State Commun.* **9**, 1949 (1971).
51. M. Ball, M. J. M. Leask, W. P. Wolf, and A. F. G. Wyatt, *J. Appl. Phys.* **34**, 1104 (1963).
52. J. C. Norvell, W. P. Wolf, L. M. Corliss, J. M. Hastings, and R. Nathans, *Phys. Rev.* **186**, 557 (1969).
53. J. C. Norvell, W. P. Wolf, L. M. Corliss, J. M. Hastings, and R. Nathans, *Phys. Rev.* **186**, 567 (1969).
54. G. A. Baker, Jr., *Phys. Rev.* **129**, 99 (1963).
55. M. F. Sykes, D. L. Hunter, D. S. McKenzie, and B. R. Heap, *J. Phys. A: Gen. Phys.* **5**, 667 (1972).
56. J. W. Stout and E. Catalano, *J. Chem. Phys.* **23**, 2013 (1955).

56

57. C. Domb and A. R. Miedema, *Progress in low Temperature Physics, Vol. 4,* edited by C. J. Gorter (North-Holland, Amsterdam, 1964).

58. G. K. Wertheim and D. N. E. Buchanan, *Phys. Rev.* **161**, 478 (1967).

59. Y. Shapira, *Phys. Rev. B* **2**, 2725 (1970).

60. M. A. Nielsen and I. L. Chuang, *Quantum Computation and Quantum Information* (Cambridge University Press, Cambridge, 2000).

61. M. Nakahara and T. Ohmi, *Quantum Computing: From Linear Algebra to Physical Realizations* (Taylor & Francis, London, 2008).

62. D. G. Cory, A. F. Fahmy, and T. F. Havel, *Proc. Natl. Acad. Sci. USA* **94**, 1634 (1997).

63. D. G. Cory, M. D. Price, W. Maas, E. Knill, R. Laflamme, W. H. Zurek, T. F. Havel, and S. S. Somaroo, *Phys. Rev. Lett.* **81**, 2152 (1998).

64. N. A. Gershenfeld and I. L. Chuang, *Science* **275**, 350 (1997).

65. I. L. Chuang, L. M. K. Vandersypen, X. Zhou, D. W. Leung, and S. Lloyd, *Nature* **393**, 143 (1998).

66. J. A. Jones and M. Mosca, *J. Chem. Phys.* **109**, 1648 (1998).

67. E. Knill, I. Chuang, and R. Laflamme, *Phys. Rev. A* **57**, 3348 (1998).

68. R. Laflamme, E. Knill, W. H. Zurek, P. Catasti, and S. V. S. Mariappan, *Phil. Trans. R. Soc. Lond. A* **356**, 1941 (1998).

69. J. A. Jones and M. Mosca, *Phys. Rev. Lett.* **83**, 1050 (1999).

70. M. D. Price, S. S. Somaroo, A. E. Dunlop, T. F. Havel, and D. G. Cory, *Phys. Rev. A* **60**, 2777 (1999).

71. L. M. K. Vandersypen, C. S. Yannoni, M. H. Sherwood, and I. L. Chuang, *Phys. Rev. Lett.* **83**, 3085 (1999).

72. L. M. K. Vandersypen, M. Steffen, G. Breyta, C. S. Yannoni, R. Cleve, and I. L. Chuang, *Phys. Rev. Lett.* **85**, 5452 (2000).

73. L. M. K. Vandersypen, M. Steffen, G. Breyta, C. S. Yannoni, M. H. Sherwood, and I. L. Chuang, *Nature* (London) **414**, 883 (2001).

74. M. Nakahara, Y. Kondo, K. Hata, and S. Tanimura, *Phys. Rev. A* **70**, 052319 (2004).

75. Y. Kondo, *J. Phys. Soc. Jpn.* **76**, 104004 (2007).

76. H. Suwa and S. Todo, *Phys. Rev. Lett.* **105**, 120603 (2010).

77. H. Suwa and S. Todo, *arXiv*:1106.3562.

78. R. H. Swendsen and J. S. Wang, *Phys. Rev. Lett.* **58**, 86 (1987).

79. U. Wolff, *Phys. Rev. Lett.* **62**, 361 (1989).

80. K. Hukushima and K. Nemoto, *J. Phys. Soc. Jpn.* **65**, 1604 (1996).

81. N. Kawashima and K. Harada, *J. Phys. Soc. Jpn.* **73**, 1379 (2004).

82. T. Nakamura, *Phys. Rev. Lett.* **101**, 210602 (2008).

83. S. Morita, S. Suzuki, and T. Nakamura, *Phys. Rev. E* **79**, 065701(R) (2009).

84. H. F. Trotter, *Proc. Am. Math. Soc.* **10**, 545 (1959).

85. M. Suzuki, *Prog. Theor. Phys.* **56**, 1454 (1976).

86. T. Kadowaki, *Ph. D thesis, Tokyo Institute of Technology* (1998).

87. K. Tanaka and T. Horiguchi, *Electronics and Communications in Japan, Part 3: Fundamental Electronic Science* **83**, 84 (2000).

88. K. Tanaka and T. Horiguchi, *Interdisciplinary Information Science* **8**, 33 (2002).

89. H. Attias, *Proceedings of the 15th Conference on Uncertainly in Artificial Intelligence* 21 (1999).

90. L. Landau, *Phys. Z. Sowjetunion* **2**, 46 (1932).

91. C. Zener, *Proc. R. Soc. London Ser. A* **137**, 696 (1932).

92. E. C. G. Stückelberg, *Helv. Phys. Acta* **5**, 369 (1932).

93. N. Rosen and C. Zener, *Phys. Rev.* **40**, 502 (1932).

94. B. K. Chakrabarti, A. Dutta, and P. Sen, *Quantum Ising Phases and Transitions in Transverse Ising Models* (Springer Verlag, Berlin, 1996).

95. G. T. Trammel, *J. Appl. Phys.* **31**, 362S (1960).

96. A. H. Cooke, D. T. Edmonds, C. B. P. Finn, and W. P. Wolf, *J. Phys. Soc. Jpn.* **17**, Suppl. B1 481 (1962).

97. J. W. Stout and R. C. Chisolm, *J. Chem. Phys.* **36**, 979 (1962).

98. V. L. Moruzzi and D. T. Teaney, *Sol. State. Comm.* **1**, 127 (1963).

99. A. Narath and J. E. Schriber, *J. Appl. Phys.* **37**, 1124 (1966).

100. R. F. Wielinga and W. J. Huiskamp, *Physica* **40**, 602 (1969).

101. W. P. Wolf, *J. Phys.* (Paris) *32 Suppl.* **C1** 26 (1971).

102. W. Wu, B. Ellman, T. F. Rosenbaum, G. Aeppli, and D. H. Reich, *Phys. Rev. Lett.* **67**, 2076 (1991).

103. W. Wu, D. Bitko, T. F. Rosenbaum, and G. Aeppli, *Phys. Rev. Lett.* **71**, 1919 (1993).

104. D. H. Reich, B. Ellman, J. Yang, T. F. Rosenbaum, G. Aeppli, and D. P. Belanger, *Phys. Rev. B* **42**, 4631 (1990).

105. T. F. Rosenbaum, *J. Phys.: Condens. Matter* **8**, 9759 (1996).

106. D. H. Reich, T. F. Rosenbaum, G. Aeppli, and H. Guggenheim, *Phys. Rev. B* **34**, 4956 (1986).

107. J. A. Mydosh, *Spin Glasses: An Experimental Introduction* (Taylor & Francis, London, 1993).

108. P. Bak, C. Tang, and K. Wiesenfeld, *Phys. Rev. Lett.* **59**, 381 (1987).

109. R. Martoňák, G. E. Santoro, and E. Tosatti, *Phys. Rev. E* **70**, 057701 (2004).

110. S. Tanaka and S. Miyashita, *J. Phys.: Condens. Matter* **19**, 145256 (2007).

111. H. Takayama and K. Hukushima, *J. Phys. Soc. Jpn.* **76**, 013702 (2007).

112. S. Tanaka and S. Miyashita, *J. Phys. Soc. Jpn.* **76**, 103001 (2007).

113. S. Miyashita, S. Tanaka, and M. Hirano, *J. Phys. Soc. Jpn.* **76**, 083001 (2007).

114. S. Tanaka and S. Miyashita, *J. Phys. Soc. Jpn.* **78**, 084002 (2009).

115. S. Tanaka and S. Miyashita, *Phys. Rev. E* **81**, 051138 (2010), *Virtual Journal of Quantum Information* **10**, (2010).

116. S. Tanaka and R. Tamura, *J. Phys.: Conf. Ser.* **320**, 012025 (2011).

117. H. Nishimori and G. Ortiz, *Elements of Phase Transitions and Critical Phenomena* (Oxford Univ Press, Oxford, 2010).

118. N. Ito, M. Taiji, and M. Suzuki, *J. Phys. Soc. Jpn.* **56**, 4218 (1987).

119. T. W. B. Kibble, *J. Phys. A* **9**, 1387 (1976).

120. T. W. B. Kibble, *Phys. Rep.* **67**, 183 (1980).

121. W. H. Zurek, *Nature* (London) **317**, 505 (1985).

122. B. Damski, *Phys. Rev. Lett.* **95**, 035701 (2005).

123. W. H. Zurek, U. Dorner, and P. Zoller, *Phys. Rev. Lett.* **95**, 105701 (2005).

58

124. V. M. H. Ruutu, V. B. Eltsov, A. J. Gill, T. W. B. Kibble, M. Krusius, Y. G. Makhlin, B. Placais, G. E. Volovik, and W. Xu, *Nature* (London) **382**, 334 (1996).

125. V. B. Eltsov, T. W. B. Kibble, M. Krusius, V. M. H. Ruutu, and G. E. Volovik, *Phys. Rev. Lett.* **85**, 4739 (2000).

126. H. Saito, Y. Kawaguchi, and M. Ueda, *Phys. Rev. A* **76**, 043613 (2007).

127. C. N. Weiler, T. W. Neely, D. R. Scherer, A. S. Bradley, M. J. Davis, and B. P. Anderson, *Nature* (London) **455**, 948 (2008).

128. S. Suzuki, *J. Stat. Mech.* P03032 (2009).

129. S. Suzuki, *J. Phys.: Conf. Ser.* **302**, 012046 (2011).

130. R. J. Glauber, *J. Math. Phys.* **4**, 294 (1963).

131. R. Shankar and G. Murthy, *Phys. Rev. B* **36**, 536 (1987).

132. D. S. Fisher, *Phys. Rev. B* **51**, 6411 (1995).

133. J. Dziarmaga, *Phys. Rev. B* **74**, 064416 (2006).

134. G. Biroli, L. F. Cugliandolo, and A. Sicilia, *Phys. Rev. E* **81**, 050101(R) (2010).

135. G. Toulouse, *Commun. Phys.* (London) **2**, 115 (1977).

136. R. Liebmann, *Statistical Mechanics of Periodic Frustrated Ising Systems* (Springer-Verlag, Berlin/Heidelberg, GmbH, Heidelberg, 1986).

137. H. Kawamura, *J. Phys.: Condens. Matter* **10**, 4707 (1998).

138. H. T. Diep (ed.), *Frustrated Spin Systems* (World Scientific, Singapore, 2005).

139. S. Tanaka and S. Miyashita, *Prog. Theor. Phys. Suppl.* **157**, 34 (2005).

140. S. Tanaka and S. Miyashita, *J. Phys. Soc. Jpn.* **76**, 103001 (2007).

141. R. Tamura and N. Kawashima, *J. Phys. Soc. Jpn.* **77**, 103002 (2008).

142. S. Tanaka and S. Miyashita, *J. Phys. Soc. Jpn.* **78**, 084002 (2009).

143. R. Tamura and N. Kawashima, *J. Phys. Soc. Jpn.* **80**, 074008 (2011).

144. R. Tamura, N. Kawashima, T. Yamamoto, C. Tassel, and H. Kageyama, *Phys. Rev. B* **84**, 214408 (2011).

145. K. Husimi and I. Syozi, *Prog. Theor. Phys.* **5**, 177 (1950).

146. R. M. F. Houtappel, *Physica* **16**, 425 (1950).

147. G. H. Wannier, *Phys. Rev.* **79**, 357 (1950).

148. G. H. Wannier, *Phys. Rev. B* **7**, 5017 (1973).

149. K. Kano and S. Naya, *Prog. Theor. Phys.* **10**, 158 (1953).

150. Y. Matsuda, H. Nishimori, and H. G. Katzgraber, *J. Phys.: Conf. Ser.* **143**, 012003 (2009).

151. Y. Matsuda, H. Nishimori, and H. G. Katzgraber, *New J. Phys.* **11**, 073021 (2009).

152. S. Tanaka, M. Hirano, and S. Miyashita, *Lecture Note in Physics "Quantum Quenching, Annealing, and Computation"* (Springer) **802**, 215 (2010).

153. S. Tanaka, to appear in *proceedings of Kinki University Quantum Computing Series: "Symposium on Quantum Information and Quantum Computing"* (2011).

154. S. Tanaka and R. Tamura, *in preparation*.

155. E. H. Fradkin and T. P. Eggarter, *Phys. Rev. A* **14**, 495 (1976).

156. S. Miyashita, *Prog. Theor. Phys.* **69**, 714 (1983).

157. H. Kitatani, S. Miyashita, and M. Suzuki, *Phys. Lett.* **108A**, 45 (1985).
158. H. Kitatani, S. Miyashita, and M. Suzuki, *J. Phys. Soc. Jpn.* **55**, 865 (1986).
159. P. Azaria, H. T. Diep, and H. Giacomini, *Phys. Rev. Lett.* **59**, 1629 (1987).
160. S. Miyashita and E. Vincent, *Eur. Phys. J. B* **22**, 203 (2001).
161. R. Tamura, S. Tanaka, and N. Kawashima, *Prog. Theor. Phys.* **124**, 381 (2010).
162. S. Tanaka and R. Tamura, and N. Kawashima, *J. Phys.: Conf. Ser.* **297**, 012022 (2011).

SPIN GLASS
A BRIDGE BETWEEN QUANTUM COMPUTATION AND
STATISTICAL MECHANICS

MASAYUKI OHZEKI

Department of Systems Science, Graduate School of Informatics, Kyoto University,
Yoshida-Honmachi, Sakyo-ku, Kyoto 606-8501, Japan
E-mail: mohzeki@i.kyoto-u.ac.jp
http://www-adsys.sys.i.kyoto-u.ac.jp/mohzeki/

In this chapter, we show two fascinating topics lying between quantum information processing and statistical mechanics. First, we introduce an elaborated technique, the surface code, to prepare the particular quantum state with robustness against decoherence. Interestingly, the theoretical limitation of the surface code, accuracy threshold, to restore the quantum state has a close connection with the problem on the phase transition in a special model known as spin glasses, which is one of the most active researches in statistical mechanics. The phase transition in spin glasses is an intractable problem, since we must strive many-body system with complicated interactions with change of their signs depending on the distance between spins. Fortunately, recent progress in spin-glass theory enables us to predict the precise location of the critical point, at which the phase transition occurs. It means that statistical mechanics is available for revealing one of the most interesting parts in quantum information processing. We show how to import the special tool in statistical mechanics into the problem on the accuracy threshold in quantum computation.

Second, we show another interesting technique to employ quantum nature, quantum annealing. The purpose of quantum annealing is to search for the most favored solution of a multivariable function, namely optimization problem. The most typical instance is the traveling salesman problem to find the minimum tour while visiting all the cities. In quantum annealing, we introduce quantum fluctuation to drive a particular system with the artificial Hamiltonian, in which the ground state represents the optimal solution of the specific problem we desire to solve. Induction of the quantum fluctuation gives rise to the quantum tunneling effect, which allows nontrivial hopping from state to state. We then sketch a strategy to control the quantum fluctuation efficiently reaching the ground state. Such a generic framework is called quantum annealing. The most typical instance is quantum adiabatic computation based on the adiabatic theorem. The quantum adiabatic computation as discussed in the other chapter, unfortunately, has a crucial bottleneck for a part of the optimization problems. We here introduce several recent trials to overcome such a weakpoint by use of developments in statistical mechanics.

64

Through both of the topics, we would shed light on the birth of the interdisciplinary field between quantum mechanics and statistical mechanics.

Keywords: Spin glass; phase transition; topological error correcting code; quantum annealing

1. Introduction: Statistical Mechanics and Quantum Mechanics

Quantum mechanics is a method partially based on the concept of the probability in outputs of the measurements on the physical state. This is because the physical state is allowed to be expressed by superposition of the possible eigenstates with the probability amplitude, which can be determined by the Schrödinger equation. Readers might feel uneasy since such a probabilistic phenomena seems to be problematic and uncontrollable. However, as shown in this chapter, several techniques have been designed and proposed for realization of the stability in quantum systems.

The quantum state is fragile, sensitive to noisy environment effects and continues to change. This peculiar destruction of the quantum state is known as decoherence. If we control to maintain the original state, we have to cure the quantum state suffering from undesired errors due to decoherence. The technique to remove these errors is the quantum error correction, which occupies the first part of this chapter. To fix successfully the quantum state, we need to prepare the redundant degrees of freedom to restore the original property and check where and how errors exist. We must thus design and deal with a many-body quantum system, which is known as an intractable problem. However it is worthwhile to strive such a difficult problem. Once you obtain an ingenious way to keep the quantum state, you find a pavement toward the realization of the quantum information processing, since you can encode simultaneously multiple information in the single quantum state by use of superposition. We here introduce a quantum error correcting technique, surface code,[1] which skillfully use the concept of the topology to encode the original state in quantum many-body system. The surface code has a remarkable feature closely related with the main concept of this chapter. The accuracy threshold, which represents the theoretical limitation to restore the original quantum state by the surface code, is related with the phase transition in the special model in spin glasses.[2] Spin glass is a disordered magnetic material, which shows a peculiar behavior with extraordinary slow relaxation toward equilibrium state in a low temperature.[3-7] The complicated interactions in spin glasses spoil several methods used for systematic analyses in statistical mechanics. The identification of the critical point in spin glasses is used to be an intractable problem for long days. Fortunately, the recent development of theory in spin glasses yields a systematic analysis for its precise location. It means that the specialized tool to analyses in spin glasses is available for the problem in quantum computation. We show the fascinating connection between two unrelated topics, while reviewing the systematic approach to derive the accuracy threshold.

By using such a technique to maintain the quantum state as introduced above, we can prepare a coherent state. Then, how do we use superposition of the quantum system? Superposition enables us to evolve the quantum system in a parallel way over the various possible states. A famous algorithm given by Shor[8] is also based on the simultaneous search over all the possible candidates. In the second part of the present chapter, we introduce a generic technique by use of quantum nature in this chapter, called quantum annealing.[9–16] Quantum annealing is an active use of superposition in order to obtain the most important result by searching all the candidates. Its attractive feature is simplicity and generality in applications. This is useful to solve several particular issues known as the optimization problems, in which we desire to find a minimizer or maximizer of the given multivariable function.[17,18] The Shor's algorithm uses a skillful technique in order to efficiently provide a desired answer. On the other hand, quantum annealing basically takes a simpler way only by controlling quantum fluctuations. The most typical procedure of quantum annealing, quantum adiabatic computation, is superior to its classical counterpart, simulated annealing,[19,20] owing to usage of quantum fluctuation. However a bottleneck of quantum adiabatic computation is also revealed in the application to the particular optimization problems. This weak point can be understood thorough a special mapping from the classical stochastic time evolution to the particular quantum dynamics. We are again between statistical mechanics and quantum information processing. Therefore, in this chapter, we will go further to overcome the bottleneck of quantum adiabatic computation by use of this fascinating relationship. We show several attempts for surmounting the obstacle involved in quantum adiabatic computation by employing several alternative strategies from statistical mechanics.

Statistical mechanics uses the probability similarly to quantum mechanics, but is capable to predict a definite behavior in the future in a large-number of components called as thermodynamic limit. The mathematical background of statistical mechanics, large deviation, provides the ability to give a definite answer even by use of the probability. Observant readers can find out the key point in this chapter. That is to deal with a large number of components. Turning on our eyes on substances around our daily life, they shows stable appearance. Material consists of many components, atoms, molecules and their mixtures, which should follow the rules of the quantum mechanics. The stability of these macroscopic systems comes from the particular property of the collection of many components. However, in many body systems, once if you tune the external parameters as temperature, pressure, and some fields, they can eventually change their uniform appearances. For instance, increase of temperature makes change of solid into liquid and gas. This phenomena is known as the phase transition. The phase transition involves singularities in the behavior of the physical quantities. The both of quantum error correction and quantum annealing shown in this chapter, suffer from this peculiar behavior. The failure to keep the quantum state by the surface code and the decay of the performance in quantum adiabatic computation come from the properties of the phase transition. Therefore, by dealing with the problems on the phase transition, we study the fascinating interdisciplinary connection between statical mechanics and quantum information processing.

Recent developments in manufacturing and manipulation of the quantum mechanical system enables us to prepare a large number of components. In this sense, the topics we deal with in this chapter must be valuable for developments in actual applications in the future.

2. Training: Statistical Mechanics

For unfamiliar readers with statistical mechanics and many students to understand the essential parts in this chapter, let us get back to the starting point of statistical mechanics.

At first let us recall the concept of the probability. What was the probability? If you have a non-tricky coin, you can say that the probability you can find a face (head or tail) should be a half. What does it mean?

2.1. *Student's misreading point: Probability is...*

We use an artificial variable with a binary $S = \pm 1$ to represent the coin state as head $(+1)$ or tail (-1). A non-tricky coin takes the probability as $P_1(S = 1) = 1/2$ and $P_1(S = -1) = 1/2$. We also can deal with a tricky coin with a biased probability as

$$P_1(S) = \frac{\exp(KS)}{Z_1(\beta)}, \tag{1}$$

where K is the strength of bias, and $Z_1(K)$ is a normalized constant explicitly given as $2\cosh K$. By use of the above probability, we can evaluate two characteristic quantities of the probabilistic system. The first quantity is expectation expressed as m, which indicates tendency of the coin whether head or tail,

$$m = \sum_{S=\pm 1} SP_1(S) = \tanh(K). \tag{2}$$

It readily shows that the coin state tends to be head if we increase the bias of the tricky coin $K \to \infty$. Another is the variance defined through the square of the difference from the expectation as

$$\sigma^2 = \sum_S (S - m)^2 P_1(S) = 1 - \tanh^2(K). \tag{3}$$

If we increase the bias K, the variance can vanish. However this is the case you know that this coin must be head without any probabilistic factors because $P_1(S = -1) \to 0$.

Then, if you once glance at the result of the single trial, can you judge whether its behavior is closely related with the obtained value of m or not? The answer is "No". The value of m only gives "tendency". The actual results will fluctuate around the expectation in repeating observations. We must repeat the coin games many times.

The above is the usual statement around the explanation of the probability. Since students, who studied the probability theory or other related topics, do not find out its genuine meaning, they often say that "probability" is obscure, abstract, and difficult!!, and conclude that it is not a well-understandable concept. Unfortunately, quantum mechanics and statistical mechanics with the concept of probability might be considered as non-reliable and non-deterministic to predict the future-coming behavior of the system we deal with. I would like to ask you why the probability for a non-tricky coin was a half. The students must recognize what the expectation express in our future. As you know, the expectation is not relevant for understanding the output in a single coin game. Let us play the game repeatedly. Then...

2.2. Probability describes... a certain behavior

In order to clarify the concept of the probability, in particular expectation, let us consider to accumulate the results of N-time observations after flipping coins. In all the trials, their results are characterized by the individually and independent distribution, which is same as in Eq. (1). Gradually approaching the specific example of use of probability in statistical mechanics, we take a simple model of the magnetic material, which is the typical example of studies in this field.

Let us introduce the Ising variable, which describes the magnetic momentum of spins in the magnetic material, often simply termed as "spin" or "Ising spin". The Ising spin takes only two integers as ± 1. We here regard the behavior of N-independent spins as the flipping coins in the N-sequential times. In statistical mechanics, we choose the specific probability distribution depending on the conditions of the system under consideration in order to calculate the expectation. Notice that, in statistical mechanics, the probability distribution can be given "$a\ priori$" similarly to the above case of a non-tricky coin. In the previous simple case, we could choose a half as the probability for the coin since it was reasonable. Generally we do not know the precise structure of the probability distribution for various events. On the other hand, statistical mechanics provides us with the probability distribution of the behavior in equilibrium state. For instance, for the equilibrium state without change of the number of components, we use the canonical distribution characterized by the Hamiltonian and temperature.

$$P(S_1, S_2, \cdots, S_N) = \frac{1}{Z_N(\beta)} \exp\left(-\beta H(S_1, S_2, \cdots, S_N)\right), \qquad (4)$$

where we define the inverse temperature $\beta = 1/k_B T$ (often $k_B = 1$) and $Z_N(\beta)$ is called as the partition function. The Hamiltonian describes the energy depending on the microscopic state of the system. The temperature controls thermal fluctuation to drive microscopic degrees of freedom in the system. If $\beta \to 0$, the spins are not stable due to strong effect by thermal fluctuation. On the other hand, in the low temperature $\beta \to \infty$, the system is settled into a lower energy state. The single spin favors parallel direction to the magnetic field since its energy can be described by the Hamiltonian $H(S) = -hS$, where h is the strength of the magnetic field. Then the joint probability of the N spins (N-time coin games) can be written by the canonical distribution as

$$P_N(S_1, S_2, \cdots, S_N) = \frac{1}{Z_N(\beta)} \prod_{i=1}^{N} \exp(\beta h S_i), \qquad (5)$$

where

$$Z_N(\beta) = \left(2\cosh(\beta h)\right)^N. \qquad (6)$$

The partition function plays a roll of the characteristic function. Its logarithmic function is called the free energy $-\beta F_N(\beta) = \log Z_N(\beta)$. We often use the thermal average by use of the canonical distribution in statistical mechanics written as

$$\langle O \rangle = \sum_{\{S_i\}} O(\{S_i\}) P_N(S, S_2, \cdots, S_N). \tag{7}$$

where O denotes the physical observable.

Let me ask the following question. If you have huge number of the outputs in an experiment, how do you interpret the result. Usually we use the average over all the realizations to simplify the outputs. This simplified picture is often called as coarse graining in statistical mechanics. What we must take care of is that the average (sample mean) over all the realizations

$$m_N = \frac{1}{N} \sum_{i=1}^{N} S_i, \tag{8}$$

and its variance

$$\sigma_N = \frac{1}{N} \sum_{i=1}^{N} (S_i - m_N)^2. \tag{9}$$

Statistical mechanics gives a definite future of the above average for the many-body system in terms of the expectation and variance estimated from the specific distribution function.

Below let us see the connection between quantities given by the empirical measure and those by the *a priori* distribution function.

2.3. Large deviation property

We do not care the detail on the system with the large-number components. For instance, for the above N spin systems, it is enough to characterize the magnetic property by the average of the N Ising spins. This is called as the magnetization, which indicates the magnetic strength of the material. Statistical mechanics predicts such significant quantities in the macroscopic scale by averages over all the components, coarse-grained quantities. If we consider the large-number limit $N \to \infty$, the averaged values can correspond to the expectations estimated by the probability distribution under a certain condition shown below. It is convenient to change the expression of the joint probability (5) into the form of the probability of the average as

$$P_N(m) = \sum_{\{S_i\}} \delta \left(m - \frac{1}{N} \sum_{i=1}^{N} S_i \right) P_N(S_1, S_2, \cdots, S_N). \tag{10}$$

By use of the integral form of the delta function as

$$\delta(x) = \int_{-i\infty}^{i\infty} d\tilde{x} \exp(x\tilde{x}), \tag{11}$$

we can obtain

$$P_N(m) \propto \sum_{\{S_i\}} \int d\tilde{m} \exp\left\{ Nm\tilde{m} + \left(-\tilde{m} \sum_{i=1}^{N} S_i + \beta h \sum_{i=1}^{N} S_i \right) \right\}. \tag{12}$$

Before performing the integration, we sum over the spin variables.

$$P_N(m) \propto \int d\tilde{m} \exp\left\{ Nm\tilde{m} + N \log\left(2\cosh\left(\beta h - \tilde{m}\right)\right) \right\}. \tag{13}$$

The integrand is proportional to N. In that case, we can apply the saddle-point method to the above integration. When we consider the infinite limit of N, the integral is given by the maximizer $\tilde{m}^*$ of the integrand as

$$P_N(m) = \exp\left\{ -Nf(m, \tilde{m}^*) \right\}. \tag{14}$$

Here we define the pseudo free energy $f(m, \tilde{m})$ as

$$f(m, \tilde{m}) = -m\tilde{m} - \log\left(2\cosh\left(\beta h - \tilde{m}\right)\right). \tag{15}$$

The saddle-point equation for $\tilde{m}$, which gives the maximum of the integrand (or the minimum of the pseudo free energy), is

$$\frac{\partial f}{\partial \tilde{m}} = 0 \to m = \tanh\left(\beta h - \tilde{m}\right). \tag{16}$$

The maximizer of the integrand is given by $\tilde{m}^* = \beta h - \tanh^{-1}(m)$. By use of this maximizer, we obtain the probability of the magnetization as

$$P_N(m) = \exp\left\{ N\left(m\beta h - m\tanh^{-1} m + \log\frac{2}{\sqrt{1-m^2}} \right) \right\}. \tag{17}$$

In Fig. 1, we describe the behavior of $f(m, \tilde{m}^*)$. The maximum is located at $m^* = \tanh(\beta h)$, which can be verified by derivative with respect to m. If N increase, the probability away from $m^* = \tanh(\beta h)$ decrease exponentially. The maximum indicates the value of the expectation since

$$\langle m \rangle = \left\langle \frac{1}{N} \sum_{i=1}^{N} S_i \right\rangle = \frac{1}{N} \frac{\partial P_N(m)}{\partial \hat{m}} = \tanh(\beta h). \tag{18}$$

Moreover, if we evaluate the variance, we reach

$$\langle \sigma_N^2 \rangle = \left\langle \frac{1}{N} \sum_{i=1}^{N} (S_i \right\rangle = \frac{1}{N} \frac{\partial^2 P_N(m)}{\partial \hat{m}^2} = \frac{1}{N} \left(1 - \tanh^2(\beta h) \right). \tag{19}$$

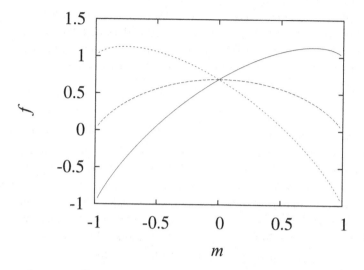

Fig. 1. Behavior of $f(m, \tilde{m}^*)$ with $\beta h = 1.0$ (solid), 0 (thick symmetric dashed curve), and -1.0 (dotted curve).

Therefore the magnetization can take a "definite" value $m^* = \tanh(\beta h)$ if $N \to \infty$. This property is known as the large deviation principle in mathematics. In the above simple case, we treat the individually-independent distribution. Fortunately, it is confirmed that the large deviation principle holds in relatively wide varieties rather than individually-independent distribution. Statistical mechanics plentifully uses explicitly and implicitly this fruitful property. It means that, in many-body system, we can predict a definite future even by use of the probability as classical mechanics with the deterministic equation. In other words, statistical mechanics predicts the average value for the physical observables of the system with large number components from the calculation of the expectation by use of an *a priori* distribution function such that the canonical, micro canonical, and ground canonical ensemble.

2.4. Mean-field analysis

The probability can give a definite future expressed by the expectation value, if we are interested in many-body systems by the support of the large deviation principle. However many degrees of freedom in material strongly interact with each other. The spin is not an exception. Differently from the above simplified case, we must take care of the effect through magnetic-dipole interactions to more precisely understand the behavior of magnetic material. We exemplify its simplest model to deal with effects of interactions known as the Ising model, whose Hamiltonian is

$$H = -J \sum_{\langle ij \rangle} S_i S_j. \tag{20}$$

The subscript stands for the index of the site, where the spin is located. The summation is taken over the nearest-neighboring pairs of spins. The geometric property as the locations of the spins is closely related with the structure of the magnetic material. As precisely as possible we analyze an actual behavior of magnetic material, we must consider a finite-dimensional structure of atoms. However it is difficult to straightforwardly perform non-trivial analyses on the finite-dimensional Ising model. We take several approximations or numerical simulations for obtaining meaningful results of the finite-dimensional Ising models. The exactly solvable examples without any approximations are found in one and two dimensions but rare exceptions.

The simple but qualitative approximation often taken as the first trial is the mean-field analysis. It enables us to obtain a meaningful picture to describe a peculiar behavior of the many-body system. The probability distribution of the degrees of freedom should be correlated with each other in the case of the existence of the interactions. However we simply assume that the probability distribution should be factorized as individually-independent one as in Eq. (5). As seen above, under this approximation, we can characterize the system by use of the magnetization m. In other words, the surrounding spins adjacent to a particular spin being replaced by a uniform variable m, we convert the many-body problem into a simple one-body system. Let us rewrite the Hamiltonian by the replacement of the surrounding spins $S_j = m$ to S_i as

$$H_{\mathrm{MF}}(S_i) = -\frac{zJm}{2} \sum_{i=1}^{N} S_i, \tag{21}$$

where z is the coordination number (usually in the case of the hyper cubic system, $z = 2d$, where d is the dimension). Similar calculations to the previous example give rise to the final expression of the probability for m as

$$P_N(m) \propto \int d\tilde{m} \exp \left\{ Nm\tilde{m} + N \log \left(2 \cosh \left(\frac{\beta Jz}{2} m - \tilde{m} \right) \right) \right\}. \tag{22}$$

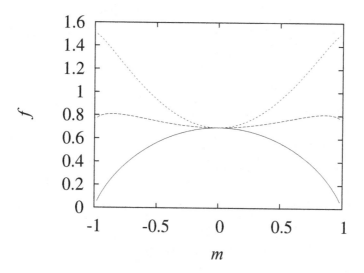

Fig. 2. Behavior of $f(m, \tilde{m}^*)$ for $z = 6$, while we set $J = 1$, with $\beta = 0.0$ (solid), 0.1 (dashed curve), and 0.25 (dotted curve).

The saddle point equation is

$$\frac{\partial f}{\partial \tilde{m}} = 0 \rightarrow m = \tanh\left(\beta \frac{Jzm}{2} - \tilde{m}\right). \tag{23}$$

Thus we find the probability of the magnetization for the interacting system

$$P_N(m) \propto \exp\left\{N\left(\frac{\beta Jz}{2}m^2 - m\tanh^{-1}m - \log\frac{2}{\sqrt{1-m^2}}\right)\right\}. \tag{24}$$

We describe the behavior of the free energy as in Fig. 2 Readers can find two maximizers in the low-temperature region. Let us evaluate explicitly this unexpected result. The most probable state is given by the maximizer of the following self-consistent equation

$$m = \tanh(\beta Jzm). \tag{25}$$

The explicit value of the magnetization over all temperature is shown in Fig. 3. Beyond the special point $\beta = 1/zJ$, the magnetization suddenly take non-zero value despite of absence of magnetic field. This is the spontaneous symmetry breaking, in other terms, the phase transition. This is also a peculiar property in many-body systems. The mean-field analysis gives several meaningful properties including the phase transition on many-body systems as above. The results by the mean-field analysis can be validated in

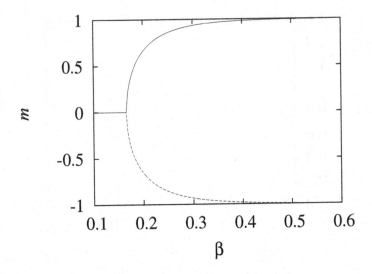

Fig. 3. Magnetization given by the mean-field analysis for the Ising model $z = 6$. Here we set $J = 1$.

infinite dimensions and reflect on the behavior in such higher dimensions. Therefore we must take care of their applicability to the actual behavior of the many-body system in finite dimensions.

2.5. Phase transition

Let us more qualitatively discuss the phase transition through the results obtained by mean-field analysis. Without the magnetic field, the spins do not have any tendency of the direction, that is $m \approx 0$. Indeed the most probable state is given by $m = 0$. This solution describes the paramagnetic phase of the magnetic material without any magnetization. However if you expand the self-consistent equation under the assumption that m should be small as

$$m = \beta J z m. \tag{26}$$

We find the special temperature $\beta = 1/zJ$, at which the magnetization m can take a non-zero value. This is a signature of the phase transition from the paramagnetic state into the ferromagnetic state. We expand the self-consistent equation up to third order of m, and thus obtain

$$m = \beta J z \left\{ m - \frac{1}{3} \left(\beta J z\right)^2 m^3 \right\}. \tag{27}$$

This equation indeed gives the ferromagnetic solutions with non-zero magnetization as $m = \pm\sqrt{3(1 - \beta Jz)}/(\beta Jz)$, which vanish at $\beta = 1/zJ$. This special temperature is called the critical point. In addition, two phases, paramagnetic and ferromagnetic phases are characterized by the behavior of the magnetization. Such a quantity, which can distinguish the phase, is called as the order parameter. Could you find the fact that you have two ferromagnetic solutions of the magnetization? It implies that the two maximizers, say two probable states, are allowed to appear in Eq. (24). Then, which solution should be chosen and will appear naturally in experience? It depends on the dynamics rule and the initial conditions. If you design the dynamics to simulate thermal fluctuations At least, we can find the fact that, in the ferromagnetic phase, state with $m = 0$ would not be expected to occur in the case with a large N. That means that two of the probable states are separate and can not change one into another in the actual experience. In other words, the system has some robustness against thermal fluctuations. That will be an analogous property to protect the fragile quantum state later.

2.6. *Spin glasses*

The above exemplified model have a uniform interactions between the Ising spins. The theoretical model of spin glasses is usually not the case. The simplest model, the Edwards-Anderson model, is defined by the following Hamiltonian

$$H = -\sum_{\langle ij \rangle} J_{ij} S_i S_j, \tag{28}$$

where J_{ij} is the disordered interactions assumed to obey several type of the distribution functions. We often deal with two typical cases, which take a bimodal distribution (then called as $\pm J$ Ising model)

$$P(J_{ij}) = p\delta(J_{ij} - J) + (1 - p)\delta(J + J_{ij}), \tag{29}$$

and the Gaussian distribution with the average J_0 and the variance J

$$P(J_{ij}) = \frac{1}{\sqrt{2\pi J^2}} \exp\left(-\frac{1}{2J^2}(J_{ij} - J_0)^2\right). \tag{30}$$

The partition function should be dependent of the specific configuration as

$$Z(\beta; \{J_{ij}\}) = \sum_{\{S_i\}} \prod_{\langle ij \rangle} \exp\left(\beta J_{ij} S_i S_j\right). \tag{31}$$

In order to evaluate the partition function and thus the free energy, we strive the difficult task to deal with the non-uniform interactions. Instead, we usually take a wise strategy to evaluate the averaged free energy based on the self-averaging property as, in the large-limit N,

$$\frac{1}{N}F(\beta; \{J_{ij}\}) \to \frac{1}{N}[F(\beta; \{J_{ij}\})], \tag{32}$$

where the square bracket denotes the average over all the combinations of $\{J_{ij}\}$ (configurational average) and the free energy is defined as

$$-\beta F(\beta; \{J_{ij}\}) = \log Z(\beta; \{J_{ij}\}). \tag{33}$$

The self-averaging property is valid for other observables, which can be obtained from the free energy per site. For instance, the magnetization per site satisfies

$$m = \frac{1}{N}\sum_{i=1}^{N} S_i = \langle S_i \rangle \to [\langle S_i \rangle]. \tag{34}$$

The average through the logarithmic term is still the intractable task. The replica method is then useful to perform the configurational average for the free energy. First, we evaluate the averaged power of the partition function to the natural number n as $[Z^n(\beta; \{J_{ij}\})]$. Then we consider the analytical continuation of n and take the limit based on the elementary identity as

$$[\log Z(\beta; \{J_{ij}\})] = \lim_{n \to 0} \frac{[Z^n(\beta; \{J_{ij}\})] - 1}{n}. \tag{35}$$

You see the replica method everywhere in analyses on spin glasses.

We avoid the complicated details of analytical results in spin glasses in order to straightforwardly understand the most important parts in this chapter. However we write down a few things on spin glasses related with our topics. The peculiar feature in spin glasses is the extraordinary slow relaxation toward equilibrium state in the low temperature. This is because the existence of many minima of the free energy on spin glasses revealed by the mean-field analysis. This fact implies that there are a large number of the most probable states. The number diverges exponentially as increase of the number of spins N. Unfortunately, this fact has been confirmed by the mean-field analysis. We have still not given the answer whether the actual finite-dimensional spin glasses follow the same scenario as that given by the mean-field analysis or not. The main reason of lack of understanding of finite dimensional spin glasses is absence of systematic tools to approach the issue. One of the exceptional methods would be the gauge theory.

2.7. Gauge theory

We take the $\pm J$ Ising model as an instance of the spin glass to show the detailed analysis. For simplicity, we combine the strength interaction with the inverse temperature as $K = \beta J$. We write the Hamiltonian of the $\pm J$ Ising model as

$$H = -\sum_{\langle ij \rangle} J\tau_{ij} S_i S_j. \tag{36}$$

The sign of the interactions follow the distribution function

$$P(\tau_{ij}) = p\delta(\tau_{ij} - 1) + (1 - p)\delta(\tau_{ij} + 1). \tag{37}$$

The partition function can be also written as $Z(K; \{\tau_{ij}\})$.

We then define the local transformation by a binary variable $\sigma_i = \pm 1$, called as the gauge transformation, as[7,22]

$$\tau_{ij} \to \sigma_i \sigma_j \tau_{ij} \tag{38}$$

$$S_i \to \sigma_i S_i. \tag{39}$$

Notice that this transformation does not change the value of the physical quantity given by the average over τ_{ij} and S_i since it alters only the order of the summations. The Hamiltonian can not change its form after the gauge transformation since the right-hand side is evaluated as

$$-\sum_{\langle ij \rangle} J\tau_{ij} \sigma_i \sigma_j \sigma_i S_i \sigma_j S_j = H. \tag{40}$$

As this case, if the physical quantity is invariant under the gauge transformation (gauge invariant), we can evaluate its exact value even for finite-dimensional spin glasses. The key point of the analysis by the gauge transformation is on the form of the distribution function. Before performing the gauge transformation, the distribution function can take the following form as

$$P(\tau_{ij}) = \frac{e^{K_p \tau_{ij}}}{2\cosh K_p}, \tag{41}$$

where $\exp(-2K_p) = (1-p)/p$. The gauge transformation changes this form, in a different way from the Hamiltonian, as

$$P(\tau_{ij}) = \frac{e^{K_p \tau_{ij} \sigma_i \sigma_j}}{2\cosh K_p}. \tag{42}$$

Let us evaluate the internal energy by aid of the gauge transformation here. The thermal average of the observables O is defined as, by use of the canonical distribution

$$\langle O \rangle_K = \sum_{\{S_i\}} \frac{1}{Z(K; \{\tau_{ij}\})} O \prod_{\langle ij \rangle} \exp(K\tau_{ij} S_i S_j). \tag{43}$$

Then the thermal average of the Hamiltonian can be written as

$$\langle H \rangle_K = \sum_{\{S_i\}} \frac{1}{Z(K; \{\tau_{ij}\})} H \prod_{\langle ij \rangle} \exp(K\tau_{ij} S_i S_j) \tag{44}$$

$$= -J \frac{d}{dK} \log Z(K; \{\tau_{ij}\}). \tag{45}$$

We can use the self-averaging property here since this is given by the derivative of the free energy, and thus take the configurational average as

$$[\langle H \rangle_K]_{K_p} = \sum_{\{\tau_{ij}\}} \prod_{\langle ij \rangle} \frac{\exp(K_p \tau_{ij})}{2 \cosh K_p} \times \langle H \rangle_K, \tag{46}$$

where $[\cdots]_{K_p}$ denotes the configurational average with K_p. Then we perform the gauge transformation, which does not change the value of the internal energy

$$[\langle H \rangle_K]_{K_p} = \sum_{\{\tau_{ij}\}} \prod_{\langle ij \rangle} \frac{\exp(K_p \tau_{ij} \sigma_i \sigma_j)}{2 \cosh K_p} \times \langle H \rangle_K. \tag{47}$$

Therefore we here take the summation over all the possible configurations of $\{\sigma_i\}$ and divide it by 2^N (the number of configurations) as

$$[\langle H \rangle_K]_{K_p} = \frac{1}{2^N} \sum_{\{\sigma_i\}} \sum_{\{\tau_{ij}\}} \prod_{\langle ij \rangle} \frac{\exp(K_p \tau_{ij} \sigma_i \sigma_j)}{2 \cosh K_p} \times \langle H \rangle_K. \tag{48}$$

We take the summation over $\{\sigma_i\}$ in advance of that over $\{\tau_{ij}\}$ and then find the partition function with K_p instead of K.

$$[\langle H_K \rangle]_{K_p} = \frac{1}{2^N} \sum_{\{\tau_{ij}\}} \frac{Z(K_p; \{\tau_{ij}\})}{(2 \cosh K_p)^{N_B}} \times \langle H \rangle_K. \tag{49}$$

Going back to Eq. (44), we can delete both of the partition functions on the denominator and numerator by setting $K_p = K$ as

$$[\langle H \rangle_K]_K = \frac{-J}{2^N (2 \cosh K_p)^{N_B}} \sum_{\{S_i\}} \sum_{\{\tau_{ij}\}} \frac{d}{dK} \exp(K\tau_{ij} S_i S_j)$$

$$= -N_B \tanh K. \tag{50}$$

Similarly, we can evaluate the rigorous upper bound on the specific heat as well as the restriction on the structure of the phase diagram. The condition

$K_p = K$ defines the special subspace in which we can perform the exact analysis even for finite-dimensional spin glasses. This subspace is called as the Nishimori line.[7,22] On this subspace, we can reveal several rigorous properties on the structure of the phase diagram even for spin glasses by relatively simple calculations. For instance, let us consider to evaluate the local magnetization $\langle S_i \rangle$ by the gauge transformation. The local magnetization identifies the existence of the ferromagnetic order and its definition is

$$\langle S_i \rangle_K = \sum_{\{S_i\}} S_i \prod_{\langle ij \rangle} \frac{\exp(K\tau_{ij}S_iS_j)}{Z(K; \{\tau_{ij}\})}. \tag{51}$$

Let us consider to evaluate its configurational average

$$[\langle S_i \rangle_K]_{K_p} = \sum_{\{\tau_{ij}\}} \prod_{\langle ij \rangle} \frac{\exp(K_p\tau_{ij})}{2\cosh K_p} \times \langle S_i \rangle_K. \tag{52}$$

After the gauge transformation, we obtain

$$[\langle S_i \rangle_K]_{K_p} = \sum_{\{\tau_{ij}\}} \sigma_i \prod_{\langle ij \rangle} \frac{\exp(K_p\tau_{ij}\sigma_i\sigma_j)}{2\cosh K_p} \times \langle S_i \rangle_K. \tag{53}$$

By summing over all the possible configurations of the gauge variables and dividing the resulting equality by 2^N, we reach

$$[\langle S_i \rangle_K]_{K_p} = \sum_{\{\tau_{ij}\}} \frac{Z(K_p; \tau_{ij})}{2^N (2\cosh K_p)^N} \times \langle \sigma_i \rangle_{K_p} \times \langle S_i \rangle_K. \tag{54}$$

On the other hand, let us evaluate the configurational average of the product of the correlation functions with different temperatures as

$$[\langle \sigma_i \rangle_{K_p} \langle S_i \rangle_K]_{K_p} = \sum_{\{\tau_{ij}\}} \prod_{\langle ij \rangle} \frac{\exp(K_p\tau_{ij})}{2\cosh K_p} \times \langle \sigma_i \rangle_{K_p} \langle S_i \rangle_K. \tag{55}$$

On this equality, the gauge transformation only changes the sign of the interactions as $\tau_{ij} \to \tau_{ij}\sigma_i'\sigma_j'$ in the exponential function. We find the following equality by summing over $\{\sigma_i'\}$ and multiplying $1/2^N$

$$[\langle S_i \rangle_K]_{K_p} = [\langle \sigma_i \rangle_{K_p} \langle S_i \rangle_K]_{K_p}. \tag{56}$$

Let us discuss the structure of the phase diagram of the $\pm J$ Ising model by use of this equality. Setting $K = K_p$, we obtain

$$[\langle S_i \rangle_{K_p}]_{K_p} = [\langle S_i \rangle_{K_p}^2]_{K_p}. \tag{57}$$

The quantity on the right-hand side is the order parameter of the spin-glass phase, called as the spin-glass parameter. In the spin-glass phase, the directions of the spins orient are random in space. Thus m should be zero and it implies

$$m = \frac{1}{N} \sum_i S_i = \langle S_i \rangle \to \left[\langle S_i \rangle_{K_p} \right]_{K_p} \to 0. \tag{58}$$

On the other hand, if the spins are frozen, their square should be non-zero. Thus

$$q = \left[\langle S_i \rangle_{K_p}^2 \right]_{K_p} \neq 0. \tag{59}$$

Therefore, on the Nishimori line, we prove that the spin-glass phase does not exist since the important fact $q = m$ is given by Eq. (57). In addition, if we take the absolute value of the magnetization away from the Nishimori line, we find

$$\left| \left[\langle S_i \rangle_K \right]_{K_p} \right| \leq \left| \left[\langle S_i \rangle_K \right]_{K_p} \right| \times \left| \left[\langle S_i \rangle_{K_p} \right]_{K_p} \right| \leq \left| \left[\langle S_i \rangle_{K_p} \right]_{K_p} \right|. \tag{60}$$

This inequality states that the magnetization takes the largest value on the Nishimori line along the vertical line of K_p. It implies that a special critical point, multicritical point, is located at the most left-hand side of the phase diagram as in Fig. 4.

In the following sections, we will find several problems on spin glasses, which are lying between statistical mechanics and quantum information processing. Although it is not enough to understand the whole properties on spin glasses, let us first go to these fascinating parts. After showing the interesting connections, we will strive the specific problems related with the spin glasses we must solve then.

3. Quantum Error Correction: Surface Code

In the first part of this chapter, we describe a technique for the quantum error correction. In order to protect the vulnerable quantum state from decoherence, we consider a method for circumvention of the effects from decoherence. It is very important to perform the quantum information processing. We here show an elaborated technique by use of the property of the topology. Our approach, in short terms, is based on encoding a few "logical" qubits in a particular state, which is not disturbed directly by noise of the "physical" qubits. Here we call the quantum state describing superposition of the binary state 0 and 1 (in terms of informatics, bits) as qubits This strategy is analogous to the classical counterpart, in which we introduce some redundancy to restore the original state.

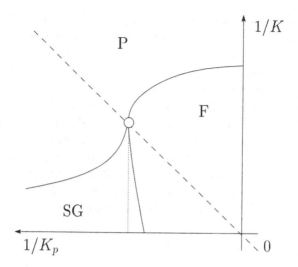

Fig. 4. Nishimori line and the typical phase diagram of the $\pm J$ Ising model. The vertical axis is the temperature, since K can be regarded as the inverse temperature. The horizontal axis expresses the density of the ferromagnetic interactions in terms of K_p. The ferromagnetic, paramagnetic, and spin-glass phases are denoted by "F", "P" and "SG", respectively. The solid curve represent the phase boundary. The dashed curve describes the Nishimori line. The multicritical point is located at the most left-hand side.

3.1. Error model

The quantum state of the single qubit is written as

$$|\Psi\rangle = \alpha|0\rangle + \beta|1\rangle. \tag{61}$$

Here the above coefficients follow $|\alpha|^2 + |\beta|^2 = 1$. We can write all the changes on the binary Hilbert space by combination of the identity operator and the Pauli operators X, Y and Z.

$$X = \begin{pmatrix} 0 & 1 \\ 1 & 0 \end{pmatrix}, \quad Y = \begin{pmatrix} 0 & i \\ -i & 0 \end{pmatrix}, \quad Z = \begin{pmatrix} 1 & 0 \\ 0 & -1 \end{pmatrix}. \tag{62}$$

By this mean, we express the error by the action of the Pauli operator. The action of X represents a phase error, and that of Z expresses a flip error. In addition, Y is a multiple error as $Y = iXZ$.

Assuming the error will occur in a stochastic manner, let us define a noise model where the qubit gets errors as

$$\rho \to p_I \rho + (p_X X \rho X + p_Y Y \rho Y + p_Z Z \rho Z). \tag{63}$$

Although we can deal with any cases of p_I, p_X, p_Y, and p_Z, Let us start from the simple case with uncorrelated between the flip and phase errors as $p_I = (1-p)^2$ and $p_X = p_Z = p$, while $p_Y = p^2$, where $0 \leq p \leq 1$. In this case, we can independently treat two of the errors. We must construct an ingenious procedure to recover the damaged qubits, that is the quantum error correction. Once an error occurs on the single qubit system, we can not remove the error, since we do not know the original state. The single qubits is too weak to save the specific information once error occurs. Therefore we need to prepare many qubits against errors as the first strategy. The first stage of the quantum error correction is thus to construct the many-body quantum system to store several bits of the information. That is analogous with the concept of redundancy in the classical information.

3.2. Surface code

Let us construct an array of qubits on a torus by setting qubits on each edge (ij) of the square lattice embedded on a torus (genus 1). Here we make two ways to describe the square lattice, while being unchanged of the location of the qubits, as in Fig. 5. The original square lattice is convenient to explain how to detect and remove the flip errors and the other (dual) is for the phase errors. The embedded qubits (say physical qubits) on all the edges actually suffer from disturbance by noisy environment. We do not directly use the quantum state of these physical qubits to encode the information. Instead, we make the particular quantum state, which is stable against direct effects due to the errors. In order to construct such a fault-tolerant subspace, called as the codespace, for encoding the information, we define two of the check operators. One is the star operator for each site s as

$$X_s = \otimes_{(ij)\in s} X_{(ij)}, \tag{64}$$

and the other is the plaquette operator for each plaquette p

$$Z_p = \otimes_{(ij)\in p} Z_{(ij)}, \tag{65}$$

where the product consists of four edges adjacent to each site and each plaquette as depicted in Fig. 6. All of the star operators X_s and the plaquette operators Z_p commute and are thus simultaneously diagonalizable. The codespace to encode the specific information consists of their simultaneous $+1$ eigenstates of all the check operators as $X_s|\Psi_c\rangle = +1|\Psi_c\rangle$ for any s and $Z_p|\Psi_c\rangle = +1|\Psi_c\rangle$ for any p. The site on the original square lattice is

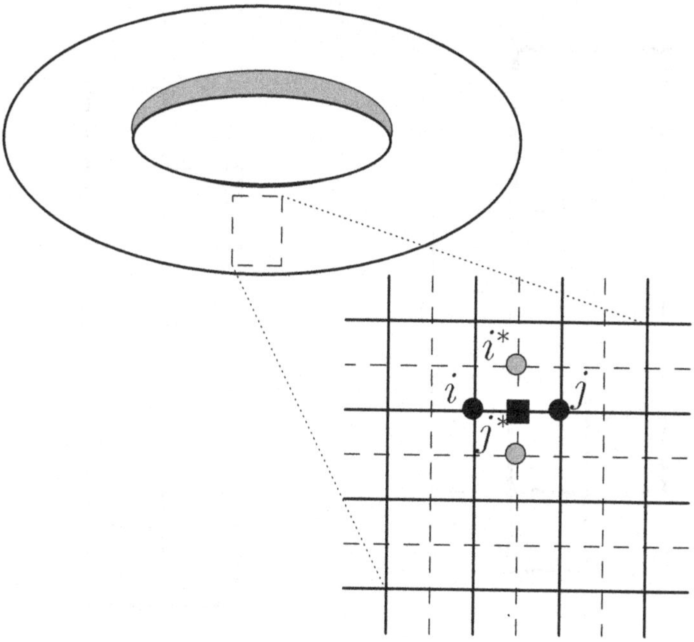

Fig. 5. Square lattice (say original) and its dual on a torus. The location of the qubit is the center of the edge (ij) (or (i^*j^*)) denoted by the black square.

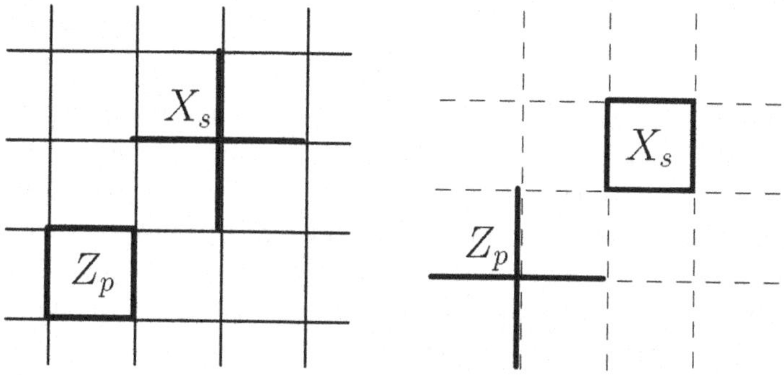

Fig. 6. Star and plaquette operators.

replaced by the plaquette on the dual one, and vice versa. The star operators are thus found to consist of the unit loops on the "dual" square lattices.

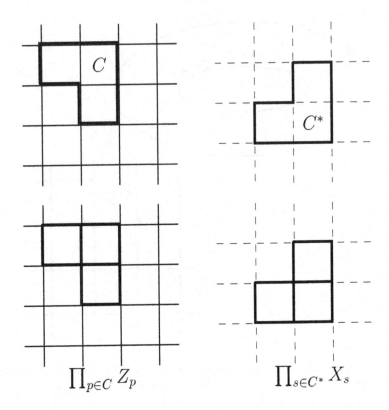

Fig. 7. Contractible loops C and C^* and their expressions by the check operators X_s and Z_p.

Let us consider contractible loops on the surface of the torus denoted by C and C^* on the original and dual lattices, respectively. The action C^* can be expressed by the product of X_s as in Fig. 7. Here notice that the double action of the same Pauli operator becomes unity. On the other hand, the plaquette operators conform the unit loop on the original square lattice. Thus the action represented by any contractible loops on the original lattice C is given by the product of Z_p as in Fig. 7. The action described by C and C^* acts trivially on the codespace, since the codespace is the eigenspace of the star and plaquette operators. In other words, the compatible-loop actions by the Pauli operator Z on the original lattice and X on the dual one can not alter the particular quantum state encoded on the torus.

We here show the explicit form of the encoded quantum state below. Let us prepare a uniform linear-combination state of all the possible contractible loops with the products of X_s and Z_p as

$$|\Psi_c\rangle = \left(1 + \sum_s X_s + \sum_{s_1,s_2} X_{s_1,s_2} + \cdots \right)$$

$$\times \left(1 + \sum_p Z_p + \sum_{p_1,p_2} Z_{p_1} Z_{p_2} + \cdots \right)|\Psi_0\rangle, \qquad (66)$$

where $|\Psi_0\rangle$ is the vacuum state of the physical qubits. Then we can easily confirm that $X_s|\Psi_c\rangle = |\Psi_c\rangle$ and $Z_p|\Psi_c\rangle = |\Psi_c\rangle$.

On the other hand, any non-contractible loop, winding around the torus, of the Pauli operators of X and Z can map the codespace to itself in a nontrivial manner, since the possible combinations of the contractible loops never constitute any non-contractible loops, while such an operator can commute with any star and plaquette operators. If we set $L \times L$ lattice on a torus, we have $2L^2$ physical qubits and $2(L^2 - 1)$ check operators. The remaining degrees of freedom of 2 implies existence of two non-contractible loops, winding around the hole of the torus L_v and winding around the body of the torus L_t as depicted in Fig. 8. Let us express these non-contractible loops on the torus in terms of the products of Pauli operators as

$$\bar{X}_v = \prod_{(ij)\in L_v} X_{(ij)} \qquad (67)$$

$$\bar{X}_t = \prod_{(ij)\in L_t} X_{(ij)} \qquad (68)$$

$$\bar{Z}_v = \prod_{(ij)\in L_v^*} Z_{(ij)} \qquad (69)$$

$$\bar{Z}_t = \prod_{(ij)\in L_t^*} Z_{(ij)}. \qquad (70)$$

They are termed as logical operators. Here we use the asterisk denoting the non-contractible loop on the dual lattice. The combinations of non-contractible loops yield $2^4 = 16$ different homology classes embedded in the original and dual square lattices on the single torus. The elementary manipulation confirms that the logical operators can form Pauli algebra of two effective qubits encoded in the topological degrees of freedom on the torus as $[\bar{Z}_v, \bar{Z}_t] = [\bar{X}_v, \bar{X}_t] = 0$, while $\bar{X}_v \bar{Z}_t = -\bar{Z}_t \bar{X}_v$ and $\bar{X}_t \bar{Z}_v = -\bar{Z}_v \bar{X}_t$. Thus, by use of these algebras by the non-contractible loops on the torus, we can prepare a 4×4 Hilbert space, say two logical qubits, explicitly given by $|\Psi_{k_1,k_2,k_3,k_4}\rangle = Z_v^{k_1} Z_t^{k_2} X_v^{k_3} X_t^{k_4} |\Psi_c\rangle$, where k_i is the number taking 0

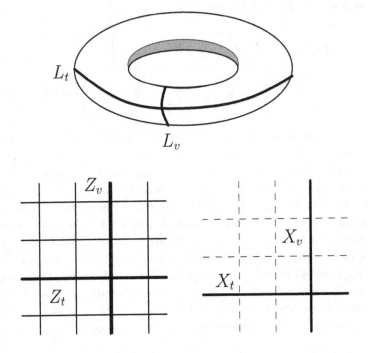

Fig. 8.   Non-contractible loops L_v and L_t and the logical operators.

and 1 to distinguish the basis of the logical qubits. The logical qubits is in general written as

$$|\Psi\rangle = \sum_{k_1,k_2,k_3,k_4} A_{k_1,k_2,k_3,k_4} |\Psi_{k_1,k_2,k_3,k_4}\rangle, \qquad (71)$$

where A_{k_1,k_2,k_3,k_4} is the coefficient following $\sum_{k_1,k_2,k_3,k_4} |A_{k_1,k_2,k_3,k_4}|^2 = 1$.

We can indeed prepare the particular quantum state from many redundant qubits as shown above. Our next interest should be how this quantum subspace is stable against the errors on the physical qubits.

3.2.1. Check operators and error syndrome

The effect coming from the decoherence, error, can occur everywhere on the torus. Let us assume that the error can be individually independently generated on each physical qubit on the torus through the channel model defined in Eq. (63). The errors $Z_{(ij)}$ and $X_{(ij)}$ can be described as, in

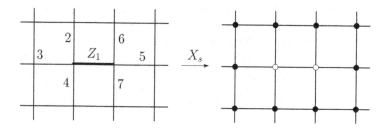

Fig. 9.   Detection of the error by the check operator. The left panel shows the original error chain E on the edge 1. The right panel describes the outputs of the star operators, which describe the endpoints of the error ∂E. The white circles denote nontrivial outputs -1 and the black ones represent $+1$.

general, non-closed error chains E and E^* on both of the original and dual lattices. We should detect the location of the errors to circumvent their effects. The star and plaquette operators can also play a roll to generate the partial information of the errors on the physical qubits. Thus these are often called the check operators. The endpoints of the error chains ∂E and ∂E^* can be detected by applications of star and plaquette operators due to anti-commutation of error with adjacent operators as follows. For instance, let us consider the local error by the action of Z on the original square lattice as in Fig. 9. The action of the error can be written as $Z_1|\Psi\rangle$. If we apply the star operator around Z_1 on the square lattice, we obtain a pair of the outputs with the negative value -1 as

$$X_1X_2X_3X_4Z_1|\Psi\rangle = -Z_1X_1X_2X_3X_4|\Psi\rangle = -Z_1|\Psi\rangle \tag{72}$$

$$X_1X_5X_6X_7Z_1|\Psi\rangle = -Z_1X_1X_5X_6X_7|\Psi\rangle = -Z_1|\Psi\rangle, \tag{73}$$

where we use anti commutation as $Z_1X_1 = -X_1Z_1$. On the other hand, the flip error by X can be detected by the plaquette operator on the dual lattice. At least, we can identify the location of the endpoints on the error chains denoted as ∂E and ∂E^*.

We do not know the exact shape of the error chains E and E^*. We must remove the effect of the errors on the torus only by use of the outputs of the check operators, error syndrome, ∂E and ∂E^*. Thus we must infer the original state from the damaged one with negative outputs on several sites and plaquettes. To remove the damage signaled by the negative eigenvalues, we consider to apply additional actions of the product of the Pauli operators. This procedure is to simply connect two endpoints of the error chains we can know. The additional actions with the original error chains can conform closed loops. The resulting effects are trivial on the codespace. However, if nontrivial, we can no longer restore the original quantum state.

In order to make the resulting effect trivial, we have to choose additional actions while inferring the same homology class as the logical operators encoding the original information only with the knowledge of the endpoints. The best way (optimal procedure) is to find out the most probable homology class. It reads, if we define $P(\bar{E}, \bar{E}^*|\partial E, \partial E^*)$ as the probability of the inferred homology class $\bar{E}$ and $\bar{E}^*$ conditioned on ∂E and ∂E^*,

$$\max_{\bar{E}, \bar{E}^*} P(\bar{E}, \bar{E}^*|\partial E, \partial E^*), \tag{74}$$

where $\bar{E}$ and $\bar{E}^*$ denote the possible homology class of the error chains. Observant readers begin to recognize an existence of the limitation of the above procedure. First, increase of the error probability p might allow the possibility of the non-contractible error chains wounding the torus. If we prepare sufficient large torus, this is not expected to occur as $\sim p^L \to 0$. As this instance, the system size L becomes larger, tolerance of the surface code against the error can be enhanced. However, if p is not small, we can not always determine the same homology class with the error chains among several candidates, since we tend to infer the wrong homology class due to the existence of long and many error chains. The connections among the original error chains and the additional actions would conform a long loop wounding the torus if p exceeds a threshold. Such an accuracy threshold actually exists and can be determined by special technique developed in the theory on the phase transition in statistical mechanics. In practice, the ideal error correction by inferring the most probable homology class of the error chains should be too harmful due to computational costs. We need to take some approximate technique in a moderate time. However, if we can, it is important to estimate the optimal error threshold in order to reveal the theoretical limitation of the surface code.

In the following, we construct the statistical-mechanical model to pave the way to analytically estimate the optimal accuracy threshold.

3.2.2. *Probability of error chains*

Notice that our task is to elucidate the macroscopic property in the error correcting procedure, not actual many-body quantum system. The outputs of the measurements of the check operators and the procedure of the application of additional actions based on the inferred homology class are classical. Although we deal with the problem on the quantum error correction, all the methods we employ here may not be quantum.

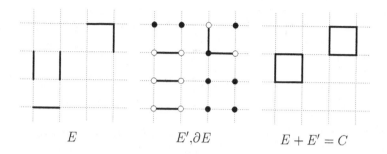

$$E \qquad\qquad E',\partial E \qquad\qquad E + E' = C$$

Fig. 10. Reasonable connections E' inferred from ∂E. For transparency, the square lattice is depicted by the dotted lines.

The error on each qubits occur in the probabilistic manner. Therefore let us construct a probabilistic model of the quantum error correction on the torus. In order to deal with the asymptotic behavior of the error pattern and inferred chains in the infinite-number limit, we use the most suited technique, statistical mechanics. Statistical mechanics starts from constructing a probabilistic model in the microscopic level. Following the concept of the large deviation principle, we take an infinite number limit of the system size in order to predict the deterministic macroscopic behavior.

Now we take the simple case with uncorrelated errors. We can separately deal with the flip and phase errors. For simplicity, hereafter we consider only the phase errors given by the actions of Z on the original square lattice.

From the knowledge of endpoints ∂E, error syndrome, we have to infer the most likely homology class of the error chains, while considering any reasonable choices.

For instance, let us consider the reasonable connections E' between each pair of the endpoints ∂E as in Fig. 10. As a result, we can make contractible loops denoted by C on the square lattice. Then the reasonable chains E' inferred from ∂E are in an equivalent class with the error chains. However there are several cases to infer non-equivalent class with the error chains as in Fig. 11.

By use of the Bayes theorem, we consider to evaluate the probability to infer the equivalent homology class with the original error chains conditioned on those endpoints as follows

$$P(\bar{E}|\partial E) = \frac{P(\partial E|\bar{E})P(\bar{E})}{\sum_{k_1,k_2,k_3,k_4} P(\partial E|\bar{E} + D_{k_1,k_2,k_3,k_4})P(\bar{E} + D_{k_1,k_2,k_3,k_4})}, \quad (75)$$

where D_{k_1,k_2,k_3,k_4} denotes the logical operator. The equivalent class is

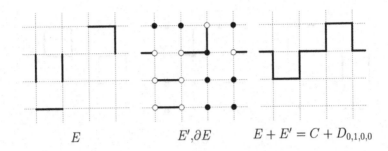

Fig. 11. Failure of inference of the equivalent class with E. The wounding chain by the product of Z yields Z_t denoted by $(k_1, k_2, k_3, k_4) = (0, 1, 0, 0)$.

given simply by $D_{0,0,0,0} = \phi$. If we infer the different class denoted by $(k_1, k_2, k_3, k_4) \neq (0, 0, 0, 0)$, the error correction fails. We assume that the prior probability for the homology classes is uniform, namely $P(\partial E) = 1/16$, since we encode various combinations of the different homology classes. Our task is to consider $P(\partial E | \bar{E})$, which is the probability to generate the endpoints of the specific error chains. Let us first evaluate the probability for the specific error chains. The error chains are generated following the independently identical distribution as

$$P(E) = \prod_{\langle ij \rangle} p^{\frac{1+\tau_{ij}^E}{2}} (1-p)^{\frac{1-\tau_{ij}^E}{2}}$$

$$= \{p(1-p)\}^{\frac{N_B}{2}} \prod_{\langle ij \rangle} \left(\frac{p}{1-p} \right)^{\frac{\tau_{ij}^E}{2}}, \tag{76}$$

where τ_{ij}^E represents the error chains and takes ± 1 ($\tau_{ij}^E < 0$, when $(ij) \in E$). Here we introduce the expression of the exponential form as in the case for the spin glass and obtain

$$P(E) = \prod_{\langle ij \rangle} \frac{\exp(K_p \tau_{ij}^E)}{2 \cosh K_p}. \tag{77}$$

Notice that K_p is minus of the original definition in the spin glass. Then, in order to evaluate $P(\partial E | \bar{E})$, we sum over all the possible configurations $E' = E + C$ of the error chains sharing the endpoints. We reach

$$P(\partial E | \bar{E}) \propto \sum_C \prod_{\langle ij \rangle} \exp(K_p \tau_{ij}^E \tau_{ij}^C), \tag{78}$$

where the summation is taken over all the possibilities of C and the product is over all the edges. We use the same indicators as τ_{ij}^E for C.

This probability tells us how to proceed the error correction. We simply perform the connection of the endpoints in the stochastic manner. If we know p in advance, it is better to set the same value. Otherwise we have

$$P_K(\partial E|\bar{E}) \propto \sum_C \prod_{\langle ij \rangle} \exp(K\tau_{ij}^E \tau_{ij}^C). \tag{79}$$

The parameter K stands for the importance/preference to choose the reasonable choice in the connection of the endpoints. For instance, in the case with $K \to \infty$, E' with the minimum length are preferred. The performance of the error correction depends on the value of K.

The loop constraints $\prod_{\langle ij \rangle} \tau_{ij}^C = 1$ allow to use another expression by the Ising variables $\tau_{\langle ij \rangle}^C = S_i S_j$. By use of this expression, we can find that $P(\bar{E})$ is written by the partition function of the $\pm J$ Ising model as

$$P_K(\partial E|\bar{E}) \propto Z(K; \{\tau_{ij}\}) = \sum_{\{S_i\}} \prod_{\langle ij \rangle} e^{K\tau_{ij}^E S_i S_j}, \tag{80}$$

where τ_{ij}^E is the sign of the random coupling in context of spin glasses. Each of the random couplings follows the distribution function of the error chains $P(E)$. Notice that, in this case, p denotes the density of the antiferromagnetic interactions. Therefore the problem to identify the error threshold of the surface code is reduced into evaluating the quantity related with the partition function of the simple spin-glass model, the $\pm J$ Ising model on the square lattice. If we increase p, the spin-glass system will be not rigid against thermal fluctuations, since the ferromagnetic order decays. Similarly, in the surface code, increase of p is expected to lead the error correction to be unfeasible. These analogous implies the existence of a fascinating relationship between the phase transition in the spin glass and the error threshold in the surface code.

In order to more clearly see this connection between statistical mechanics and quantum error correction, we show a schematic picture of the phase diagram for the $\pm J$ Ising model on the square lattice as in Fig. 12. In the low-temperature region with a relatively small p, the ferromagnetic ordered phase can be observed. It means that the order of the Ising spins exists and suppresses the fluctuation of the domain walls (boundaries between different signed spins). Let us consider to evaluate $P_K(\bar{E}|\partial E)$ in the ferromagnetic phase. We rewrite the probability as, in terms of the free energy of the spin

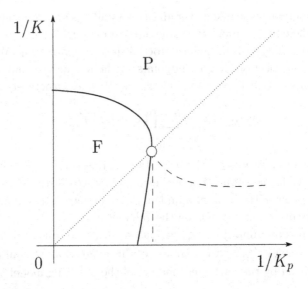

Fig. 12. Phase diagram of the $\pm J$ Ising model. The figure is depicted by the same symbols in Fig. 4. In the two-dimensional case, the absence of the spin-glass phase has been partially proved and supported by numerical simulations.[23]

glass,

$$P_K(\bar{E}|\partial E) = \frac{1}{1 + \sum_{\{k_i\}} \exp(-K\Delta F_{k_1,k_2,k_3,k_4})}. \tag{81}$$

Here the difference of the free energy $\Delta F_{k_1,k_2,k_3,k_4}$ is given as

$$\Delta F_{k_1,k_2,k_3,k_4} = F(K; \{\tau_{ij}\}) - F(K; \{\tau_{ij}\tau_{ij}^{D_{k_1,k_2,k_3,k_4}}\}), \tag{82}$$

where we use K instead of β as

$$F(K; \{\tau_{ij}\tau_{ij}^{D_{k_1,k_2,k_3,k_4}}\}) = -\frac{1}{K}\log Z(K; \{\tau_{ij}\tau_{ij}^{D_{k_1,k_2,k_3,k_4}}\}). \tag{83}$$

In context of statistical mechanics, the logical operator D_{k_1,k_2,k_3,k_4} is interpreted as the induction of the antiferromagnetic interactions along the non-contractible loop as $\{\tau_{ij}\tau_{ij}^{D_{k_1,k_2,k_3,k_4}}\}$. In the ferromagnetic phase, such a long boundary of antiferromagnetic interactions yields a crevice with the same scale as the inducted loop in the ordered spins. Therefore the free energy difference should become $\sim \mathcal{O}(L)$. As a result, the probability to infer the equivalent homology class with the original error chains is asymptotically, in the limit of $L \to \infty$

$$P_K(\bar{E}|\partial E) \to 1. \tag{84}$$

This means that the error correction is feasible.

On the other hand, in a high temperature and for not a small p, namely the paramagnetic phase, the order of the spins is destroyed. The vast number of islands of the different-signed spins can exist and the boundaries fluctuate. The free-energy difference becomes zero, and thus we reach.

$$P_K(\bar{E}|\partial E) \to 1/16. \tag{85}$$

This means that the error correction is infeasible. Therefore the locations of the critical points of the spin-glass model identify the error thresholds of the surface code. In particular, the optimal threshold is located on the special critical point along the Nishimori line, namely multicritical point,[7,22] since it is at the most right side in the phase diagram. In other words, if we know the precise value of p, it is easier to correct the errors in the logical qubits.

3.3. Analyses on accuracy thresholds for surface code

In general, the problem on finite-dimensional spin glasses is intractable. However recent development in theory of this realm enables us to estimate a precise location of the critical point in several spin glasses. In the following section, we introduce such a specialized theory in detail.

3.3.1. Duality analysis: Simple case

The situation that no systematic analytical methods attacking the problems on the critical phenomena in finite-dimensional spin glasses have been changed since a recent development in the spin-glass theory. It enables us to estimate the precise value of the special critical point especially on the Nishimori line, which corresponds to the optimal error threshold.[24-27] The method as shown below is based on the duality, which can identify the location of the critical point especially on two-dimensional spin systems.[28,29] Let us review the simple case of the Ising model.

The duality is a symmetry argument by considering the low and high-temperature expansions of the partition function. As a result, we can obtain a simple relation between two different temperatures through the partition function as

$$Z(K) = \Lambda Z(K^*), \tag{86}$$

where K^* denotes the dual coupling constant related with the original one K. The coefficient λ will be an important quantity below.

We here deal with the case for the Ising model on the square lattice, whose partition function is given as, through the Hamiltonian (20),

$$Z(K) = \sum_{\{S_i\}} \prod_{\langle ij \rangle} \exp(K S_i S_j), \tag{87}$$

where the product is taken over all the nearest neighboring pairs on the square lattice. The original formulation of the duality analysis is based on a relatively painful calculation.[28] We here show a much simpler version given by a simple Fourier transformation for the local part of the Boltzmann factor.[29]

We define the edge Boltzmann factor as

$$x_\phi(K) = \exp(K \cos \pi \phi), \tag{88}$$

where we use another binary variable as $\phi = 0$ and 1, instead of the Ising spin. In addition, we apply the binary Fourier transformation to this quantity, called as the dual edge Boltzmann factor,

$$x_k^*(K) = \frac{1}{\sqrt{2}} \sum_{\phi=0,1} x_\phi \exp(i\pi k \phi) = \frac{1}{\sqrt{2}} \left(e^K + e^{-K} \cos \pi k \right). \tag{89}$$

Then we find that the partition function can be written in two ways by both of the edge Boltzmann factors. First, the partition function is simply expressed by the original Boltzmann factor as

$$Z(K) = \sum_{\{\phi_i\}} \prod_{\langle ij \rangle} x_{\phi_{ij}}(K), \tag{90}$$

where $\phi_{ij} = \phi_i - \phi_j$. The difference between ϕ_i and ϕ_j is taken in order from left to right and from top to bottom. Inserting the inverse Fourier transformation of the dual edge Boltzmann factor yields another expression of the partition function

$$Z(K) = \left(\frac{1}{\sqrt{2}} \right)^{N_B} \sum_{\{\phi_i\}} \sum_{\{k_{ij}\}} \prod_{\langle ij \rangle} x_{k_{ij}}^*(K) e^{i k_{ij}(\phi_i - \phi_j)}. \tag{91}$$

Here let us perform the summation over $\{\phi_i\}$ by considering the adjacent edges. We concentrate on the particular site 0 and then find that the exponential term can be factorized and the summation can be taken independently as

$$\sum_{\phi_0=0,1} e^{i(-k_{10}-k_{20}+k_{03}+k_{04})\phi_0} = 2\delta(k_{10} + k_{20} - k_{03} - k_{04} \equiv 0 \ (\text{mod } 2)). \tag{92}$$

We describe adjacent edges to the particular site 0 as in Fig. 13. In order

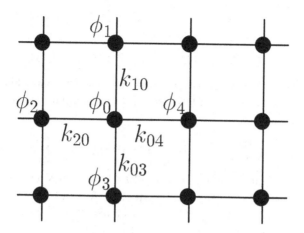

Fig. 13. Description of the terms appearing in Eq. (92)

to remove the terms of the Kronecker's delta functions, we employ another binary variable as

$$k_{10} = \phi_1^* - \phi_2^* \tag{93}$$

$$k_{20} = \phi_2^* - \phi_3^* \tag{94}$$

$$k_{03} = \phi_4^* - \phi_3^* \tag{95}$$

$$k_{04} = \phi_1^* - \phi_4^*. \tag{96}$$

We set these new variables $\{\phi_i^*\}$ on each plaquette on the square lattice, namely each site on the dual one. The resultant expression of the partition function is

$$Z(K) = \sum_{\{\phi_i^*\}} \prod_{\langle ij \rangle} x_{\phi_{ij}^*}^*(K), \tag{97}$$

where $\phi_{ij}^* \equiv \phi_i^* - \phi_j^*$. This fact leads us to the double expressions with the different arguments as

$$Z(x_0(K), x_1(K)) = Z(x_0^*(K), x_1^*(K)). \tag{98}$$

To reduce the number of the arguments, we normalize the partition function by the principal edge Boltzmann factors $x_0(K)$ and $x_0^*(K)$.

$$\{x_0(K)\}^{N_B} z(u_1(K)) = \{x_0^*(K)\}^{N_B} z(u_1^*(K)), \tag{99}$$

where z is the normalized partition function $z(u_1) = Z/\{x_0(K)\}^{N_B}$ and $z(u_1^*) = Z/\{x_0^*(K)\}^{N_B}$. We explicitly obtain $u_1(K) = x_1(K)/x_0(K) =$

$\exp(-2K)$ and $u_1^*(K) = x_1^*(K)/x_0^*(K) = \tanh K$. It reads, if we set the dual coupling as $\exp(-2K^*) = \tanh(K)$,

$$\{x_0(K)\}^{N_B} z(u_1(K)) = \{x_0^*(K)\}^{N_B} z(u_1(K^*)). \tag{100}$$

Therefore we find that the partition function for the Ising model has symmetry between two different temperatures. This is the duality. The coefficient Λ is also obtained as

$$\Lambda = \left(\frac{x_0^*(K)}{x_0(K)}\right)^{N_B} = \left\{\frac{1}{\sqrt{2}}\left(1 + \exp(-2K)\right)\right\}^{N_B}. \tag{101}$$

The well known duality relation $\exp(-2K^*) = \tanh K$ can identify the exact location of the critical point by equating $K = K^*$. The critical point is estimated as $K_c = \ln(1 + \sqrt{2})/2 = 0.440686794\cdots$. Interestingly, the coefficient Λ becomes unity at the critical point. We use this property as *a priori* assumption for the analysis on the critical point for spin glasses.

3.3.2. *Duality analysis: Spin glass*

The replica method, which is often used in theoretical studies on spin glasses, allows to generalize the duality analysis to spin glasses.[24,25] Let us consider the duality for the replicated partition function as $[Z^n(K; \{\tau_{ij}\}]$. In this case, the multiple binary Fourier transformation defines the dual edge Boltzmann factor as

$$x_{k_1,k_2,\cdots,k_n}^*(K_p, K) = \left(\frac{1}{\sqrt{2}}\right)^n \sum_{\phi_1,\phi_2,\cdots,\phi_n} x_{\phi_1,\phi_2,\cdots,\phi_n}(K_p, K)e^{i\sum_{i=1}^n k_i\phi_i}, \tag{102}$$

where the original edge Boltzmann factor is given by the configurational average of the n-replicated $\pm J$ Ising model,

$$x_{\phi_1,\phi_2,\cdots,\phi_n}(K_p, K) = \frac{1}{2\cosh K_p}\left\{e^{-K_p + K\sum_{i=1}^n \cos\pi\phi_i} + e^{K_p - K\sum_{i=1}^n \cos\pi\phi_i}\right\}. \tag{103}$$

Notice that the definition of p is the density of the antiferromagnetic interactions here to analyze the accuracy threshold for the surface code. They leads us to the double expression of the replicated partition function as, in a similar way to the above simple case,

$$\{x_0(K_p, K)\}^{N_B} z(u_1(K_p, K), u_2(K_p, K), \cdots)$$
$$= \{x_0^*(K_p, K)\}^{N_B} z(u_1^*(K_p, K), u_2^*(K_p, K), \cdots), \tag{104}$$

where the subscript of u_k and u_k^* stands for the number of anti-parallel pair among n replicas on each edge. We restrict ourselves to the case to analyze the location of the multicritical point, namely $K = K_p$ on the Nishimori line.

Unfortunately we cannot replace $u_k^*(K, K)$ by $u_k(K^*, K^*)$ even after normalization as the above case for the Ising model, since the replicated partition function is multivariable. Nevertheless we can estimate the location of the multicritical point even without the ordinary procedure of the duality. We here put a priori assumption that $x_0(K, K) = x_0^*(K, K)$ at the critical point, implying the coefficient of the double expressions of the replicated partition function should be unity. According to this assumption, we take the limit $n \to 0$ of the equation along the replica method and thus obtain

$$-p\log p - (1 - p)\log(1 - p) = \frac{1}{2}\log 2. \qquad (105)$$

The solution is $p_c = 0.1100\cdots$. We thus conclude that the accuracy threshold is estimated as $p_c = 0.1100\cdots$. Strictly speaking, this analysis is not exact. In practical, the precision of the above result has been confirmed to be satisfiable by supports from numerical simulations. In addition to such numerical validations, the following theoretical refinement of the method to identify the multicritical point have been considered.

3.3.3. Duality analysis with real-space renormalization

As above mentioned, the ordinary duality analysis hampers since the replicated partition function was multivariable. Thus we rely on the assumption that $x_0(K, K) = x_0^*(K, K)$ would give the location of the critical point. We here sketch the relationship between the double expressions of the replicated partition function as the curves of the relative Boltzmann factors $u_k(K, K)$ and $u_k^*(K, K)$ on the two-dimensional space for simplicity, although those are correctly in a hyper space as in Fig. 14. The thick curve denotes the relative Boltzmann factor $u_k(K, K)$ and the dashed one represents the dual one $u_k^*(K, K)$. When the temperature increases, the representative point of the replicated partition function moves from P (the high-temperature limit) to F (the low temperature limit) along the thick curve. On the other hand, the dual representative point goes inversely along the dashed curve. These features have been shown rigorously and imply the existence of the duality relation for the temperature[30]. If two curves become completely coincident with each other, we then obtain a relation implying $u_k^*(K, K) = u_k(K^*, K^*)$. Solving this relation, we can obtain the duality relation for different temperatures as the case of the Ising model. In spin glasses, they do not overlap.

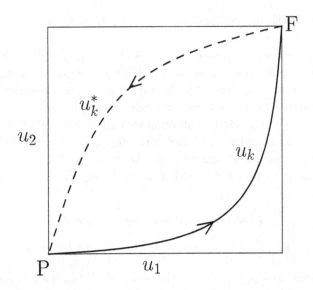

Fig. 14.   Relative Boltzmann factors of the replicated partition function (Projection onto two dimensions). The high-temperature limit is given by P, while the low-temperature limit is F. The solid curve represents the original relative Boltzmann factor. The dashed curve denotes the dual relative Boltzmann factor.

We import another piece of the theories in statistical mechanics, the real-space renormalization group analysis. Most of the problems on statistical mechanics are tractable since the degrees of freedom are highly correlated. Often we employ a trick to map the original problem into much simpler problem with a recursive structure by use of some approximation. As in Fig. 15, we trace over a part of spins on the square lattice. Repeating this procedure while omitting the generated multi-body interactions (approximation), we construct the renormalization group in the form of the recursion

$$K_n = R(K_{n-1}), \qquad (106)$$

where K_n is the renormalized coupling constant after n steps. By use of the renormalization group analysis, we describe flow of the renormalized coupling constant in K space. The flow usually terminates two fixed points representing the ordered and disordered phase, while being divided by an unstable fixed point, which is the critical point. The precision to represent the original behavior in the model is dependent on the approximation in construction of the renormalization group. We generalize this procedure

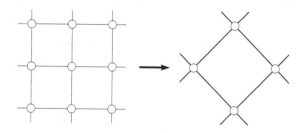

Fig. 15. Real-space renormalization. In this instance, a limited number of spins are summed, and the remaining ones construct another square lattice with multiple-body interactions.

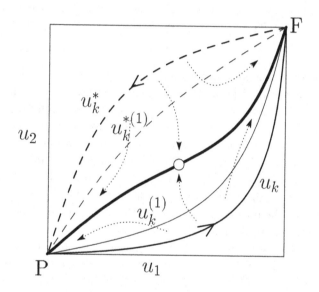

Fig. 16. Relative Boltzmann factors of the replicated partition function and their renormalization flow (Projection onto two dimensions). The dotted arrows depict the renormalization flow starting from several points on both of the relative Boltzmann factors. The internal curves denote the renormalized relative Boltzmann factors. The bold curve describes the sufficiently renormalized relative Boltzmann factors with the unstable fixed point as the case with a single variable in the partition function.

into the relative Boltzmann factors $\{u_k(K, K)\}$ and $\{u_k^*(K, K)\}$. The flow can be depicted in the hyper space of $\{u_k(K, K)\}$ as in Fig. 16. Similarly to the above simple case, the renormalization flow goes toward two fixed points P and F. The unstable fixed point C would be located between the

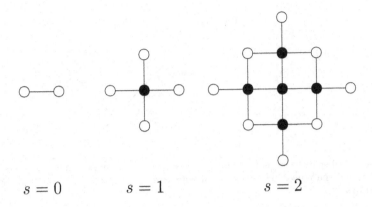

$$s = 0 \qquad\qquad s = 1 \qquad\qquad\qquad s = 2$$

Fig. 17. Clusters for duality with renormalization. The white circles denote edge spins whose configuration is represented by k, and the black ones are to be summed.

original and dual relative Boltzmann factors, since they do not overlap but have a common critical point. The renormalization flow starts from both of the relative Boltzmann factors and once moves to the unfixed point. After several renormalization steps, the flow goes to two fixed points. That means that, if we conduct one step of the renormalization for each point on the original and dual relative Boltzmann factors, we observe both of the renormalized curves get close to each other as in Fig. 16. The sufficient steps of the renormalization makes both of the curves into a common thick curve capturing the unfixed point, namely the multicritical point. As far as possible, we desire to estimate the location of the critical point precisely. Then, let us consider to trace the partial spins without any approximations. For example, in the case on the square lattice, we define the cluster Boltzmann factor $x_k^{(s)}(K, K)$, where the subscript k denotes the configuration of the edge (white-colored) spins and s expresses the size of the cluster in Fig. 17. When $k = 0$, all the edge spins are in up directions. We sum over the internal (black-colored) spins in order to perform the renormalization without approximation. It is difficult to perform the sufficient step of the renormalization without any approximation. Remember that we can estimate the relatively correct value of the multicritical point even without the renormalization. Therefore we propose a systematic way to improve precision of the location of the multicritical point. We employ the following equation to estimate the location of the multicritical point,

$$x_0^{(s)}(K, K) = x_0^{*(s)}(K, K). \tag{107}$$

The equality for $s = 0$ (edge) reproduces the case without renormalization as $p_c^{(0)} = 0.1100$.[24,25] If we increase the size of the used cluster, we can systematically approach the exact solution for the location of the multicritical point of the $\pm J$ Ising model as $p_c^{(1)} = 0.1093$ and $p_c^{(2)} = 0.1092$.

If we remove the condition of the Nishimori line $K_p = K$, we can describe the phase boundary for spin glasses by

$$x_0^{(s)}(K_p, K) = x_0^{*(s)}(K_p, K). \tag{108}$$

By use of this equation, we can obtain the precise results for the phase boundary in the higher temperature region than the Nishimori line and the whole phase boundary of the diluted Ising model.[26,27]

3.3.4. Other cases

Not only the surface code, several quantum error correcting codes are found to possess the connection with the spin-glass models. Although we omit their detailed explanations, we look over the recent results in short below. We restrict ourselves to the case in which the duality analysis have predicted the accuracy threshold.

First, we simply mention the cases by other arrangement of physical qubits in the surface code than the square lattice, say triangular and hexagonal lattices. The accuracy threshold for both of the lattices can be given as $p_c = 0.164 \cdots$ (triangular) and $p_c = 0.067 \cdots$ (hexagonal).[27] Recently, another type of the surface code with more computational capability have been developed, color code. Also in the case of the color code, we prepare the arrangement of the physical qubits on each unit triangle on the triangular lattice or Union-Jack lattice as in Fig. 18.[31,32] The color code on the Union-Jack lattice implements the whole Clifford group of unitary gates generated by the Hadamard gate, the $\{\pi/8\}$ gate, and the controlled-NOT gate, although that on the triangular lattice can not employ the $\{\pi/8\}$ gate. It means that both of the color codes have the wider computational capability than that of the Pauli group. The corresponding statistical mechanical model has three-body interactions differently from the above surface code. The quenched random interaction then represents the error arising on the physical qubits on the unit triangles. Here the duality analysis can estimate the accuracy thresholds for both of the color codes as $p_c = 0.1096 - 8$ (triangular) and $p_c = 0.1092 - 3$.[33] Interestingly, the advantage of the computational capability does not spoil the robustness of the error correction.

In addition, the practical errors in implementation of the quantum system are not only on the computational basis. Lack of the physical qubits is

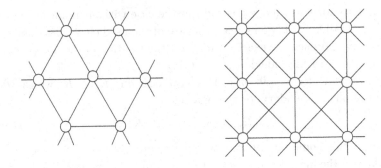

Fig. 18.   Color code on the triangular and Union-Jack lattices.

also likely to occur. Recently, Stace *et al.* have proposed a modified scheme of the surface code to protect the logical quantum state from both type of the errors, namely from decoherence and lack of the physical qubits.[34,35] Then the corresponding statistical mechanical model becomes the diluted version of the $\pm J$ Ising model. We can identify the locations of the optimal thresholds depending on the ratio of the loss of the physical qubits q as in Fig. 19.[36] The detailed values are shown in Table 1. As shown in Table 1 and Fig. 19, the decay of the robustness, depending on increase of q, of the surface code against the flip/phase errors can be observed. This behavior can be interpreted as the decay of the ferromagnetic order due to the dilution of the interactions.

Table 1.   Optimal thresholds given by the duality analyses with $s = 0$, 1, and 2 clusters for the uncorrelated case.

q	p_c $(s = 0)$	p_c $(s = 1)$	p_c $(s = 2)$
0.00	0.11003	0.10928	0.10918
0.10	0.09240	0.09196	0.09189
0.20	0.07245	0.07235	0.07233
0.30	0.04984	0.05004	0.05009
0.40	0.02462	0.02492	0.02500
0.45	0.01155	0.01174	0.01179

3.3.5. *Depolarizing channel*

Before closing this section, we add another example on the quantum error correction related with statistical mechanics. We discussed the case without

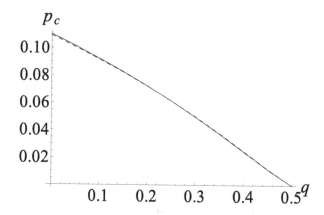

Fig. 19. Decay of the robustness due to loss of the physical qubits. The solid curve expresses the results by the duality with $s = 0$ and the dashed curve denotes those with $s = 1$, while both of them are mostly overlapping in this scale.

correlation of both of the flip and phase errors for simplicity. It is straightforward to generalize the above procedure for the depolarizing channel case such that the error probabilities are equalized as $p_X = p_Y = p_Z = p/3$. In this case, the corresponding statistical mechanical model becomes a little bit different one from the $\pm J$ Ising model. We can no longer independently deal with both of the errors. Let us first evaluate the probability for the specific error chains. The error chains are generated following the distribution function as

$$P(E, E^*) = \prod_{\langle ij \rangle} \left(\frac{p}{3}\right)^{\frac{1+\tau_{ij}^E}{4} + \frac{1+\tau_{ij}^{E^*}}{4} + \frac{1+\tau_{ij}^E \tau_{ij}^{E^*}}{4}} (1-p)^{\frac{1-\tau_{ij}^E}{4} + \frac{1-\tau_{ij}^{E^*}}{4} + \frac{1-\tau_{ij}^E \tau_{ij}^{E^*}}{4}}$$

$$\propto \left(\frac{p}{3(1-p)}\right)^{\frac{\tau_{ij}^E}{4} + \frac{\tau_{ij}^{E^*}}{4} + \frac{\tau_{ij}^E \tau_{ij}^{E^*}}{4}}. \tag{109}$$

We again use the expression of the exponential form as in the case for the spin glass and obtain

$$P(E, E^*) \propto \prod_{\langle ij \rangle} \exp(K_d \tau_{ij}^E + K_d \tau_{ij}^{E^*} + K_d \tau_{ij}^E \tau_{ij}^{E^*}). \tag{110}$$

Notice that $\exp(-4K_d) = 3(1 - p)/p$. Similarly, the problem is to identify the singularity of the following the probability through the partition function of a spin glass

$$P_K(\partial E, \partial E^* | \bar{E}, \bar{E}^*)$$

$$\propto \sum_{s_i, s_i^*} \prod_{\langle ij \rangle} \exp(K\tau_{ij}^E S_i S_j + K\tau_{ij}^{E^*} S_i^* S_j^* + K\tau_{ij}^E \tau_{ij}^{E^*} S_i S_j S_i^* S_j^*),$$

$$(111)$$

where the quenched random coupling obeys the distribution function $P(E, E^*)$. This model is regarded as a spin-glass version of the 4-state Potts model.

The duality analysis can be performed for this statistical-mechanical model in order to estimate the optimal thresholds.[37] In addition, we can apply the modified scheme to circumvent the effects due to loss of the physical qubits proposed by Stace et al.[36] We list the results for the depolarizing channel in Table 2.

Table 2. Results for the correlated case (depolarizing channel) $p_X = p_Z = p_Y = p/3$.

q	p_c $(s = 0)$	p_c $(s = 1)$
0.00	0.18929	0.18886
0.10	0.16025	0.15985
0.20	0.12690	0.12656
0.30	0.08844	0.08819
0.40	0.04454	0.04440
0.45	0.02121	0.02114

4. Quantum Annealing and Beyond

The second part of this chapter, we introduce another application of statistical mechanics in context of quantum computation. As in the other chapter, by use of the analogy of simulated annealing (SA) developed in statistical mechanics, a generic algorithm intended for solving optimization problems by use of quantum nature, quantum annealing (QA), has been studies since its proposal. In SA, we make use of thermal, classical, fluctuations to employ a stochastic search for the desired lowest-energy state, which corresponds to the optimal solution, by allowing the system to hop from state to state over intermediate energy barriers. In QA, by contrast, we introduce non-commutative operators as artificial degrees of freedom of quantum nature in order to induce quantum fluctuations. The most typical procedure of QA is performed by adiabatic control of quantum fluctuations, say quantum adiabatic computation (QAC). The adiabatic theorem guarantees that an sufficient slow time evolution would let the system closely follow the instantaneous ground state. Therefore, starting from an initial trivial ground state, we can reach a nontrivial one by slowly controlling quantum fluctuations. However, in a practical sense, we desired to perform the procedure to find the answer of the optimization problem as fast as possible. The adiabatic control to efficiently obtain the optimal solution takes a characteristic time related with the energy gap between the first excited state and the lowest one during its procedure. The problems with closure of the energy gap are difficult to be solved in a moderate time by QAC. Unfortunately, as far as we know, the typical difficult problems in classical computation involves the exponential closure of the energy gap by increase of the problem size.

In this chapter, we show several trials to overcome the bottleneck of quantum adiabatic computation by use of statistical mechanics. We use formal similarities between SA and QAC in order to import several developments in statistical mechanics. In both methods, we have to drive the system slowly and carefully to control the strengths of thermal or quantum fluctuations. The idea behind quantum adiabatic computation is to keep the system close to the instantaneous ground state of a quantum system. This is analogous to the protocol of SA, in which one tries to make the system keep quasi-equilibrium state. It is indeed possible to make this analogy more transparent by a precise formulation, from which a fascinating technique to be expected to improve the performance of QAC can be proposed.

4.1. *Quantum adiabatic computation: Short review*

Let us briefly review the procedure of the most typical QA, namely QAC, before importing theory of statistical mechanics to quantum computation. In QA, we introduce a non-commutative operator to drive the system by quantum nature as

$$H(t) = f(t)H_0 + (1 - f(t)) H_1, \tag{112}$$

where H_0 is the classical Hamiltonian consisting of diagonal elements, which express the cost function. Here $f(t)$ is assumed to be a monotonically increasing function satisfying $f(0) = 0$ and $f(T) = 1$. For instance, $f(t) = t/T$, where T denotes the computation time for QAC. The quantum annealing starts from a trivial ground state of H_1, which is often chosen to be a uniform linear combination of the computational basis as

$|\Psi(0)\rangle = |\sigma\rangle/\sqrt{N}$. Let us deal with only the discrete combinatorial optimization problems, which are simply termed as the optimization problem below. Most of the optimization problem can be expressed by the spin-glass Hamiltonian. In order to explain the procedure of QAC in detail, we consider to find the ground state of the simple spin-glass model as used before, namely the random-bond Ising model,

$$H_0 = -\sum_{\langle ij \rangle} J_{ij} \sigma_i^z \sigma_j^z, \tag{113}$$

where the summation is taken over all the nearest-neighboring pairs of the Ising spins. We take the computational basis of the eigenstates of the Ising variables to represent the instantaneous state as $|\Psi(t)\rangle = |\sigma_1^z, \sigma_2^z, \cdots, \sigma_N^z\rangle$. The transverse-field operator is often used as quantum fluctuations for implementing QAC for the spin-glass model

$$H_1 = -\Gamma_0 \sum_{i=1}^{N} \sigma_i^x, \tag{114}$$

where Γ_0 is the strength of the transverse field. The whole Hamiltonian of QAC (although widely used for QA) thus becomes

$$H(t) = f(t) \sum_{\langle ij \rangle} J_{ij} \sigma_i^z \sigma_j^z + (1 - f(t)) \Gamma_0 \sum_{i=1}^{N} \sigma_i^x. \tag{115}$$

The ground state of the transverse-field operator H_1 is trivially given by a uniform linear combination as $\sum_{\{\sigma\}} |\sigma\rangle / \sqrt{2}^N$. For a sufficiently large T, the adiabatic theorem guarantees that the instantaneous state at time t, $|\Psi(t)\rangle$, is very close to the instantaneous ground state implying $|0(t)\rangle$, $\langle 0(t)|\Psi(t)\rangle \approx 1$, when the instantaneous ground state $|0(t)\rangle$ is non-degenerate. The condition for $|0(t)\rangle$, $\langle 0(t)|\Psi(t)\rangle 1 - \epsilon^2 (\epsilon \ll 1)$ to hold is given by

$$\frac{\max \left| \langle 1(t)| \frac{dH(t)}{dt} |0(t)\rangle \right|}{\min \Delta^2(t)} = \epsilon, \tag{116}$$

where $|1(t)\rangle$ is the instantaneous first excited state, and $\Delta(t)$ is the energy gap between the ground state and first excited one. The maximum and minimum are evaluated between 0 and T. In our case, since $dH(t)/dt \propto 1/T$, the adiabatic condition is written as

$$T \propto \frac{1}{\epsilon \min \Delta^2(t)}. \tag{117}$$

Therefore, if we desire to solve the problems involved with the exponential closure of the energy gap while increase of N, QAC must take extremely long time to find the ground state with high probability.[38,39]

4.2. Novel type of quantum annealing

In order to overcome this problematic bottleneck of QAC, we change our strategy from the adiabatic control of quantum fluctuations. We demand an important key for nonequilibrium statistical mechanics by using a fascinating bridge between quantum computation as QAC and statistical mechanics. To make the connection more clear, we show a useful technique to relate both of the fields.

4.2.1. Classical quantum mapping

Let us compare two of the procedures of SA and QAC. Both of the protocols are given by slow sweep of thermal and quantum fluctuations to keep the system trace the instantaneous stationary state, equilibrium for SA and ground state for QAC, respectively. In numerical implementation of SA, we demonstrate a stochastic dynamics driven by thermal fluctuation with the master equation.

$$\frac{d}{dt}P(\sigma;t) = \sum_{\sigma'} M(\sigma|\sigma';t)P(\sigma';t), \tag{118}$$

where $P(\sigma;t)$ is the probability with a spin configuration of $\{\sigma_i^z\}$ simply denoted as σ at time t. Notice that σ is not a state of a single spin but is a collection of spin states. $M(\sigma'|\sigma;t)$ expresses the transition matrix following the conservation of probability $\sum_{\sigma} M(\sigma|\sigma';t) = 1$ and the detailed balance condition

$$M(\sigma|\sigma';t)P_{\mathrm{eq}}(\sigma';t) = M(\sigma'|\sigma;t)P_{\mathrm{eq}}(\sigma;t), \tag{119}$$

where we denote the instantaneous equilibrium distribution as $P_{\mathrm{eq}}(\sigma;t) = \exp(-\beta(t)E(\sigma;t))/Z(t)$ and the instantaneous energy $E(\sigma;t)$ is the value of the classical Hamiltonian $H_0(t)$. Since SA consists of a dynamic control of the temperature, the time variable t has been written explicitly in the arguments of the inverse temperature and the partition function. In order to satisfy this condition, we often use the transition matrix with Metropolis rule as

$$M(\sigma|\sigma';t) = \min(1, \exp(-\beta\Delta E(\sigma|\sigma';t))), \tag{120}$$

where

$$\Delta E(\sigma|\sigma';t) = E(\sigma;t) - E(\sigma';t), \tag{121}$$

or heat-bath rule as

$$M(\sigma|\sigma';t) = \delta_1(\sigma,\sigma') \frac{\exp\left(-\frac{\beta}{2}\Delta E(\sigma|\sigma';t)\right)}{2\cosh\left(\frac{\beta}{2}\Delta E(\sigma|\sigma';t)\right)}, \tag{122}$$

where

$$\delta_1(\sigma|\sigma') = \delta(2, \sum_{i=1}^{N}(1 - \sigma_i\sigma_i')). \tag{123}$$

On the other hand, the dynamics of QAC is governed by the Shrodinger equation. To look at the connection between SA and QAC, we employ a mapping technique of dynamics of relaxation toward equilibrium as the master equation into the Shrodinger equation. If we use the following special quantum Hamiltonian, we find it possible to simulate the dynamics of SA in quantum manner,[40]

$$H_q(\sigma'|\sigma;t) = \delta(\sigma',\sigma) - e^{\beta(t)H_0(\sigma')/2}M(\sigma'|\sigma;t)e^{-\beta(t)H_0(\sigma)/2}. \tag{124}$$

Here we consider to gradually increase the inverse temperature following the spirit of SA as $\beta(t)$. This special Hamiltonian has the following state as its ground state,

$$|\Psi_{\text{eq}}(t)\rangle = \frac{1}{\sqrt{Z(t)}} \sum_{\sigma} e^{-\frac{\beta(t)}{2}H_0(\sigma)}|\sigma\rangle. \tag{125}$$

It is easy to confirm that the quantum expectation value of a physical quantity $A(\sigma)$ in the ground state as $\langle\Psi_{\text{eq}}(t)|A(\sigma)|\Psi_{\text{eq}}(t)\rangle$ coincides with the thermal expectation of the same quantity with $\beta(t)$. The ground state energy simply takes zero. This fact is shown by use of the conservation of the probability and the detailed-balance condition as

$$\left(\delta(\sigma',\sigma) - e^{\frac{\beta(t)}{2}H_0(\sigma')}M(\sigma'|\sigma;t)e^{-\frac{\beta(t)}{2}H_0(\sigma)}\right)|\Psi_{\text{eq}}(t)\rangle$$

$$\propto \sum_{\sigma}\left(e^{\frac{\beta(t)}{2}H_0(\sigma')} - e^{\frac{\beta(t)}{2}H_0(\sigma')}M(\sigma'|\sigma;t)\right)|\sigma\rangle = 0. \tag{126}$$

On the other hand, the excited states have positive-definite eigenvalues, which can be confirmed by the application of the Perron-Frobenius theorem. By using the above special quantum system, we can treat a quasi-equilibrium stochastic process in SA as an adiabatic dynamics as in QAC.

The above formulation is a generic way of the classical-quantum mapping. We demonstrate the above mapping of SA into QAC by more explicit instance. Let us consider an optimization problem that can be expressed as a classical Hamiltonian with local interaction

$$H_0 = -\sum_j H_j, \tag{127}$$

where H_j involves σ_j^z and a finite number of $\sigma_k^z (k_n eqj)$. Taking a familiar instance is a spin-glass system with nearest-neighbor interactions,

$$H_j = -\sum_{k \in \partial j} J_{jk}\sigma_j^z\sigma_k^z, \tag{128}$$

where ∂j denote sites adjacent to j. The following Hamiltonian is the explicit form, which facilitates our analysis,

$$H_q^{SG}(t) = -\chi(t)\sum_j \left(\sigma_j^x - e^{\frac{\beta(t)}{2}H_j}\right), \tag{129}$$

where $\chi(t) = e^{\beta(t)p}$ with $p = \max_i|H_i|$. Notice that p is proportional to the interaction energy and is the order of $\mathcal{O}(N^0)$ due to finiteness of the interaction range. Let us consider that the protocol of SA in the above quantum system. In the first stage with very high temperature $\beta(0) \to 0$, the quantum system (129) reduces

$$H_q^{SG}(t) = -\sum_j \left(\sigma_j^x - 1\right). \tag{130}$$

Its ground state is the uniform linear combination of all possible states in the basis to diagonalize $\{\sigma_i^z\}$. It means that all states appear with an equal probability as the equilibrium distribution in high-temperature limit. Therefore the quantum system (129) can correctly demonstrate the initial condition of SA. In addition, in the limit $\beta(t \to T) \to \infty$, (129) becomes

$$H_q^{SG}(t) \sim \chi(t)\sum_j e^{\frac{\beta(t)}{2}H_j}. \tag{131}$$

The ground state is with the lowest value of the classical system (127), because each H_j takes its lowest value. These observations confirm that the quantum system (129) indeed demonstrate the quasi-stationary dynamics in SA. Notice that, Interestingly, the adiabatic condition of the above special quantum system we used in the classical-quantum mapping can reproduce the condition of convergence of SA. By use of the fascinating connection between SA and QAC, we can import several theories of statistical mechanics into the quantum dynamics.

As shown later, the collaboration of statistical mechanics with quantum dynamics can produce a new algorithm to solve optimization problems in different manners from QAC.

4.2.2. Jarzynski equality

Among several recent developments in statistical mechanics, we take the Jarzynski equality (JE) as an attempt to improve the performance of QAC.[41] This equality relates quantities at two different thermal equilibrium states with those of nonequilibrium processes connecting these two states. It can also be termed as a generalization of the well-known inequality, the second law of thermodynamics $\Delta F \leq \langle W \rangle_{0 \to T}$. Here the brackets $\langle \cdots \rangle_{0 \to T}$ are for the average taken over nonequilibrium processes between the initial (at $t = 0$) and final states (at $t = T$), which are specified only macroscopically and thus there can be a number of microscopic realizations.

The Jarzynski equality is written as[42,43]

$$\langle e^{-\beta W} \rangle_{0 \to T} = \frac{Z_T}{Z_0}. \tag{132}$$

Here the partition functions for the initial and final Hamiltonians are expressed as Z_0 and Z_T, respectively. One of the important features is that JE holds independently of the pre-determined schedule of the nonequilibrium process. Another celebrated benefit is that JE reproduces the second law of thermodynamics by using the Jensen inequality. Notice that we have to take all fluctuations into account in evaluation of the expectation value in the right-hand side of JE in order to calculate the free energy difference. The Jarzynski equality holds formally in the case with change of temperature,

$$\langle e^{-Y} \rangle_{0 \to T} = \frac{Z_T}{Z_0}, \tag{133}$$

when we employ the pseudo work instead of the ordinary performed work due to the energy difference as

$$Y(\sigma; t_k) = (\beta_{k+1} - \beta_k) E(\sigma; t_i). \tag{134}$$

Here we employ discrete time expressions as $t_0 = 0$ and $t_n = T$ for simplicity. We show the simple proof of JE for the dynamics in SA. Let us consider a thermal nonequilibrium process in a finite-time schedule governed by the master equation. The left-hand side of JE is explicitly written as

$$\langle e^{-Y} \rangle_{0 \to T} = \sum_{\{\sigma_k\}} \prod_{k=0}^{n-1} \left\{ e^{-Y(\sigma_{k+1}; t_k)} e^{\delta t M(\sigma_{k+1} | \sigma_k; t_k)} \right\} P_{\text{eq}}(\sigma_0; t_0). \tag{135}$$

We take the first product of the above equation as,

$$\sum_{\sigma_0} \left\{ e^{-Y(\sigma_1;t_0)} e^{\delta t M_0(\sigma_1|\sigma_0;t_0)} \right\} P_{eq}(\sigma_0;t_0)$$

$$= P_{eq}(\sigma_1;t_1) \frac{Z_1}{Z_0}. \tag{136}$$

Repetition of the above manipulation in Eq. (135) yields the quantity in the right-hand side of JE as,

$$\sum_{\sigma_n} P_{eq}(\sigma_n;t_n) \prod_{k=0}^{n-1} \frac{Z_{k+1}}{Z_k} = \frac{Z_n}{Z_0}, \tag{137}$$

where $Z_n = Z_T$. This is the case for a classical system on a heat bath, not for a quantum system. Although readers may think the above proof is not relevant to improvement of QAC, the formulation of JE for the classical system is available for QAC by aid of the classical-quantum mapping above introduced.

4.2.3. Quantum Jarzynski annealing

We here provide a novel protocol from the same spirit as JE by using the special quantum system (129). Let us consider to start from the trivial ground state with the uniform linear combination similarly to the case of the ordinary QA. This initial state expresses the high-temperature equilibrium state as $|\Psi_{eq}(t_0)\rangle \propto e^{-\beta(t_0)H_0(\sigma)/2}|\sigma\rangle$ with $\beta(t_0) \ll 1$. We introduce the exponentiated pseudo work operator $R(t_k) = \exp(-Y(\sigma_k;t_k)/2)$. Observant readers might think it as a non-unitary operator, but we can construct this operation by considering an enlarged quantum system as detailed later. When we apply $R(t_k)$ to $|\Psi_{eq}(t_k)\rangle$ with the inverse temperature $\beta(t_k)$, the quantum state is changed into a state corresponding to the equilibrium distribution with $\beta(t_{k+1})$. Then the application of the time-evolution operator $U(\sigma'|\sigma;t_{k+1}) = \exp(-i\delta t H_q(\sigma'|\sigma;t_{k+1})/\hbar)$ does not alter the instantaneous quantum state, since it is the ground state of $H_q(\sigma'|\sigma;t_{k+1})$. The resultant state after the repetition of the above procedure is

$$|\Psi(t_n)\rangle \propto \prod_{k=1}^{n} \left\{ R(t_k)U_k(\sigma_k|\sigma_{k-1};t_k) \right\} |\Psi_{eq}(t_0)\rangle. \tag{138}$$

The product in the right-hand side is essentially of the same form as in Eq. (135). Instead of the exponentiated matrix of $\delta t M(\sigma_{k+1}|\sigma_k;t_k)$, we use the time-evolution operator $U_k(\sigma_k|\sigma_{k-1}; t_k)$ here. After the system reaches the state $|\Psi(t_n)\rangle$, we measure the obtained state by the projection onto a specified state σ'. The probability is then given by $|\langle\sigma'|\Psi(t_n)\rangle|^2$, which means that the desired ground state can be obtained with the probability proportional to $\exp(-\beta(t_n)H_0)$, since $|\Psi(t_n)\rangle \propto |\Psi_{eq}(t_n)\rangle$. If the above procedure continues up to $\beta(t_n) \gg 1$, the resultant wave function can yield the ground state of H_0. We call this procedure as the quantum Jarzynski annealing (QJA).

We here emphasize the following three points. First, the protocol of QJA does not rely on the quantum adiabatic control. The computational time does not depend on the energy gap. In this sense, QJA does not suffer from the energy-gap closure differently from QAC. The required computational cost for realization of QJA is estimated from the number of the unitary gates as will be discussed below. Second, from the property of JE, the result is independent of the schedule to tune the parameter, T, in the above manipulations. Third, we do not need the repetition of the pre-determined process to deal with all fluctuations in the nonequilibrium-process average as in the ordinary JE, since the classical ensemble is mapped to the quantum wave function. In addition, if we obtain the final wave function, the output can give the ground state we desire with a very high probability since $\beta(t_n) \gg 1$.[a]

4.2.4. Problems in measurement of answer

So far, so good. No problems seem to exist in the realization and performance of QJA. Unfortunately, we can find a serious problem to efficiently solve the hard optimization problem by QJA. In order to implement QJA, we must prepare a peculiar operator, the exponentiated pseudo work operator $R(t_k) = \exp(-Y(\sigma; t_k)/2)$, which looks like a non-unitary operator. We can realize this non-unitary operator for the original Hilbert space by preparing an enlarged quantum system with an ancilla qubit (another two-level quantum system) as $|\Psi, \phi_1\rangle = |\Psi\rangle \otimes |\phi_1\rangle$, where ϕ_1 is assumed to take 0 and 1.[44] We call the ancila qubit the computational state below. We initially set $|\Psi, \phi_1 = 0\rangle$. For simplicity, we restrict ourselves to the case with

---

[a]Notice that several-time repetitions of experiments should be demanded since the output by quantum measurement is probabilistic. However we should emphasize that this point is not related with the theoretical property of JE attributed to rare events, necessity of all the realizations during the nonequilibrium process, but it comes from quantum nature.

$H_0(\sigma) > 0$ for any states. Let us introduce the following unitary operator for the enlarged quantum system as

$$R^{\text{unit.}}(t_k) = \sum_\sigma |\sigma\rangle\langle\sigma| \otimes \left(\begin{array}{cc} \sqrt{y_k(\sigma)} & \sqrt{1-y_k(\sigma)} \\ -\sqrt{1-y_k(\sigma)} & \sqrt{y_k(\sigma)} \end{array} \right)$$
$$\equiv I_\sigma \otimes Y_k, \qquad (139)$$

where $y_k(\sigma) = \exp(-Y_k(\sigma;t_k))$. We can obtain the weighted quantum system by applying this operator as $\sqrt{y_k(\sigma)}|\Psi,\phi_1 = 0\rangle$. Then we regard $R^{\text{unit.}}(t_k)$ as the exponentiated pseudo work operation $R(t_k)$ for the quantum state $|\Psi,\phi_1 = 0\rangle$. The other probability amplitudes of $R^{\text{unit.}}|\Psi,\phi_1 = 0\rangle$ are given as

$$\langle\Psi,0|R^{\text{unit.}}(t_k)|\Psi,0\rangle = \sqrt{y_k(\sigma)} \qquad (140)$$
$$\langle\Psi,1|R^{\text{unit.}}(t_k)|\Psi,0\rangle = \sqrt{1-y_k(\sigma)}. \qquad (141)$$

The output in measurement of the quantum state, $\sqrt{1-y_k(\sigma)}|\Psi,\phi_1 = 1\rangle$ is regarded as an undesired error state in our computation. Therefore we here do not employ such error states to find the ground state of the classical Hamiltonian $H_0(\sigma)$. Let us assume the uniform change of the inverse temperature $\beta(t_{k+1}) - \beta(t_k) \equiv \delta\beta$ for simplicity. In order to gain the relevant weight for obtaining the desired ground state of H_0, we must increase a parameter corresponding to the inverse temperature up to $\beta(t_n)\epsilon \sim 1$, where ϵ is the minimum energy gap of the "classical" Hamiltonian H_0 (usually given by the energy unit). Therefore the number of steps of QJA is necessary up to $n \equiv \beta(t_n)/\delta\beta \sim 1/\epsilon\delta\beta$. The computational time n is the size of the enlarged quantum state we have to prepare as $|\Psi,\phi_1,\cdots,\phi_n\rangle$. The initial state of QJA in this case is $|\Psi_{\text{eq}}(t_0),0,\cdots,0\rangle$. The other states as $|\Psi,1,0\cdots,0\rangle$, $|\Psi,0,1,\cdots,0\rangle$, etc. such that several ancilla qubits take different states from the initial ones as $\phi_i = 1$ are regarded as the error states. To gain the weight up to $\exp(-\beta(t_n)H_0)$, we perform the n-step exponentiated work operations as $I_\sigma \otimes Y_1 \otimes I_2 \otimes \cdots \otimes I_n$, $I_\sigma \otimes I_1 \otimes Y_2 \otimes I_3 \otimes \cdots \otimes I_n$, $\cdots$ and $I_\sigma \otimes I_1 \otimes \cdots \otimes I_{n-1} \otimes Y_n$, where I_j denotes the identity operation. We can then obtain the desired state $|\Psi,0,0,\cdots,0\rangle$ with the weight as $\exp(-\beta(t_n)H_0)$. The weights for the other states, the error states, are given as $(1 - \exp(-\beta(t_n)H_0))^{1/n}(\exp(-\beta(t_n)H_0))^{1-1/n}$ for $|\Psi,1,0,\cdots,0\rangle$ and $(1-\exp(-\beta(t_n)H_0))^{2/n}(\exp(-\beta(t_n)H_0))^{1-2/n}$ for $|\Psi,1,1,0,\cdots,0\rangle$ and so on.

Let us consider to obtain the meaningful outputs by the projective measurements. The desired state is simply one with $\phi_i = 0$ for any i. The probability for obtaining the state $|\sigma, 0, 0, \cdots, 0\rangle$ is evaluated as

$$p_0(\sigma) = \frac{e^{-\beta(t_n)H_0(\sigma)}}{\sum_\sigma \sum_{k=0}^n (1 - e^{-\beta(t_n)H_0(\sigma)})^{k/n} (e^{-\beta(t_n)H_0(\sigma)})^{1-k/n}} = \frac{e^{-\beta(t_n)H_0(\sigma)}}{2^N}.$$
(142)

Thus the successful result is generated with the exponentially decreasing probability for increase of the system size N. We will demand more ingenious techniques to efficiently obtain the desired outputs. This is the remaining problem on QJA.

4.3. Non-adiabatic quantum computation

In the previous section, we consider to implement the manipulation in the left-hand side of JE in the quantum computation by use of the classical-quantum mapping. On the other hand, we have another direction of QA to improve the performance beyond QAC by directly considering the nonequilibrium behavior of quantum system. Again, in order to develop the theory of QA, we rely on the property of JE but for the quantum system. Instead of the adiabatic control, we consider to repeat non-adiabatic quantum annealing (small or intermediate T) starting from a state chosen from equilibrium ensemble, not necessarily the ground state. We may not be able to easily reach the ground state of H_0 by such processes, since the system does not keep the instantaneous ground state as in the adiabatic computation. We instead need to repeat the process many times to hit the ground state. In this way, the problem of long annealing time is expected to be replaced by many repetitions of non-adiabatic (possibly quick) evolution. We call such a procedure as non-adiabatic quantum annealing (NQA).[45]

4.3.1. Jarzynski equality for quantum system

Before analysis on the detailed property of the non-adiabatic computation, we recall the Jarzynski equality but for isolated quantum system,[46,47] while assuming its application to NQA.

Let us consider to find the ground state as of the spin-glass Hamiltonian H_0 as in Eq. (127) by NQA. We prepare the dynamical quantum system following the time-dependent Hamiltonian (112). Initially we pick up a state from the canonical ensemble for $H(0) = H_1 = -\Gamma_0 \sum_i \sigma_i^x$ and then let it evolve following the time-dependent Schrödinger equation. The performed work in the isolated quantum system is given by the difference between the outputs of projective measurements of the initial and final energies, $W = E_m(T) - E_n(0)$. Here m and n denote the indices of the instantaneous eigenstates measured at the final and initial steps of NQA, $H(T)|m(T)\rangle = E_m(T)|m(T)\rangle$ and $H(0)|n(0)\rangle = E_n(0)|n(0)\rangle$, respectively. The time-evolution operator is given by the following unitary operator as

$$U_T = \mathcal{T} \exp\left(i \int_0^T dt H(t)\right), \qquad (143)$$

where $\mathcal{T}$ denotes the time ordered product. Thus we can evaluate the transition probability between the initial and final steps as

$$P_{m,n}(0 \to T) = |\langle \Psi_m(T)|U_T|\Psi_n(0)\rangle|^2. \qquad (144)$$

Therefore the path probability for the nonequilibrium process starting from the equilibrium ensemble as

$$P_{m,n}(0 \to T)\frac{\exp(-\beta E_n(0))}{Z_0(\beta; \{J_{ij}\})}, \qquad (145)$$

where we express the instantaneous partition function at each time t as $Z_t(\beta; \{J_{ij}\})$. By directly evaluating the left-hand side of JE, we reach JE for the isolated quantum system as

$$
\begin{aligned}
\langle e^{-\beta W} \rangle_{QA} &= \sum_{m,n} e^{-\beta W} P_{m,n}(0 \to T)\frac{\exp(-\beta E_n(0))}{Z_0(\beta; \{J_{ij}\})} \\
&= \sum_{m,n} \frac{e^{-\beta E_m(T)}}{Z_0(\beta; \{J_{ij}\})} P_{m,n}(0 \to T) \\
&= \sum_m \frac{e^{-\beta E_m(T)}}{Z_0(\beta; \{J_{ij}\})} \\
&= \frac{Z_T(\beta; \{J_{ij}\})}{Z_0(\beta; \{J_{ij}\})},
\end{aligned}
\qquad (146)
$$

where we used the fact that the performed work W is a classical number and

$$
\begin{aligned}
\sum_n P_{m,n}(0 \to T) &= \sum_n \langle \Psi_m(T)|U_T|\Psi_n(0)\rangle\langle \Psi_n(0)|U_T^\dagger|\Psi_m(T)\rangle \\
&= \sum_m \langle \Psi_m(T)|U_T U_T^\dagger|\Psi_m(T)\rangle = 1.
\end{aligned}
\qquad (147)
$$

If we measure the physical observable $\hat{O}_T$ at the last of the nonequilibrium process, we obtain another equation as

$$\langle \hat{O}_T e^{-\beta W} \rangle_{QA} = \langle \hat{O} \rangle_\beta \frac{Z_T(\beta; \{J_{ij}\})}{Z_0(\beta; \{J_{ij}\})}, \qquad (148)$$

where the subscript on the square brackets in the right-hand side denotes the thermal average in the last equilibrium state with the inverse temperature β.

Below we show several observations by application of JE for the isolated quantum system to implementation of NQA.

4.3.2. Performance of non-adiabatic quantum annealing

First we discuss the possibility of NQA as a solver. Let us consider to measure equilibrium quantities through NQA. The ratio of Eqs. (146) and (148) gives

$$\frac{\langle \hat{O}_T e^{-\beta W} \rangle_{QA}}{\langle e^{-\beta W} \rangle_{QA}} = \langle \hat{O} \rangle_{\beta}. \tag{149}$$

The resultant equation suggests that the thermal average under the Hamiltonian H_0 on the right-hand side can be estimated by NQA on the left-hand side. This fact may be useful in the evaluation of equilibrium average, since the left-hand side is evaluated without slow adiabatic processes. In order to investigate the property of the ground state, we tune the inverse temperature into a very large value $\beta \gg 1$. We should be careful because the average on the left-hand side involves a non-extensive quantity, the exponentiated work, whose value fluctuates significantly from process to process. The average on the left-hand side must be calculated by many trials of annealing processes. Thus, rare events with large values of the exponentiated work (i.e. $\beta|W| \gg \Gamma_0$) would contribute to the average significantly, and we have to repeat the annealing process very many times in order to reach the correct value of the average. It usually needs very many, typically exponentially many, repetitions. Thus the difficulty has not been relaxed yet in general, but the present new perspective may lead to different methods and tools than conventional ones to attack the problem.

4.4. Analyses on non-adiabatic quantum annealing

Unfortunately, we have not reached any answers on the performance of NQA. Instead We here evaluate several properties in nonequilibrium process as in NQA for the particular spin glasses. We can exactly analyze nonequilibrium behavior by combination of JE with the gauge transformation, although, in general, there are few exact results in nonequilibrium quantum dynamical system with many components.

Following the prescription of the Jarzynski equality, we consider a repetition of NQA starting from the equilibrium ensemble. Let us remember the whole Hamiltonian of QA for the typical spin glasses. The initial Hamiltonian is given only by the transverse field $H(0) = H_1$, which means a trivial uniform distribution. Consequently, as a starting point of our analyses, we write down the specialized JE to the case for NQA as

$$\langle e^{-\beta W} \rangle_{QA} = \frac{Z_T(\beta, \{J_{ij}\})}{(2 \cosh \beta \Gamma_0)^N}. \tag{150}$$

We assume that the interactions $\{J_{ij}\}$ follow the distribution function for the $\pm J$ Ising model (29), which is better to be rewritten as

$$P(J_{ij}) = \frac{\exp(\beta_p J_{ij})}{2 \cosh \beta_p J}, \tag{151}$$

where we do not use $K = \beta J$ for transparency, and $\exp(-2\beta_p J) = (1-p)/p$.

4.4.1. Gauge transformation for quantum spin systems

For several special spin glasses as the $\pm J$ Ising model, the gauge transformation is available for analyses on the dynamical property even under quantum fluctuations. The time-dependent Hamiltonian as in Eq. (115) is invariant under the following local transformation,

$$\sigma_i^x \to \sigma_i^x, \ \sigma_i^y \to \xi_i \sigma_i^y, \ \sigma_i^z \to \xi_i \sigma_i^z, \ J_{ij} \to J_{ij} \xi_i \xi_j \quad (\forall i, j), \tag{152}$$

where $\xi_i(= \pm 1)$ is called as a gauge variable. This transformation is designed to preserve the commutation relations between different components of Pauli matrix.[48] We skillfully use the gauge transformation to analyze the dynamical property of the nonequilibrium behavior in NQA.

4.4.2. Relationship between two different paths of NQA

Below, we reveal several properties inherent in NQA by the gauge transformation. Let us take the configurational average of Eq. (150) over all the realizations of $\{J_{ij}\}$ for the special case with $\beta = \beta_1$ and $\beta_p = \beta_2$ as

$$\left[\langle e^{-\beta_1 W} \rangle_{QA} \right]_{\beta_2} = \left[\frac{Z_T(\beta_1; \{J_{ij}\})}{(2 \cosh \beta_1 \Gamma_0)^N} \right]_{\beta_2}. \tag{153}$$

The right-hand side is written explicitly as

$$\left[\langle e^{-\beta_1 W} \rangle_{QA} \right]_{\beta_2} = \sum_{\{J_{ij}\}} \frac{\exp \left(\beta_2 \sum_{\langle ij \rangle} J_{ij} \right)}{(2 \cosh \beta_2 J)^{N_B}} \frac{Z_T(\beta_1; \{J_{ij}\})}{(2 \cosh \beta_1 \Gamma_0)^N}. \tag{154}$$

Let us here apply the gauge transformation and sum over all possible configurations of the gauge variables $\{\xi_i\}$. We obtain, after dividing the result by 2^N,

$$\left[\langle e^{-\beta_1 W}\rangle_{\mathrm{QA}}\right]_{\beta_2} = \sum_{\{J_{ij}\}} \frac{Z_T(\beta_2; \{J_{ij}\})Z_T(\beta_1; \{J_{ij}\})}{2^N (2\cosh\beta_2 J)^{N_B} (2\cosh\beta_1\Gamma_0)^N}. \tag{155}$$

A similar quantity of the average of the exponentiated work for the spin glass with the inverse temperature β_2 and the parameter for the quenched randomness β_1 gives

$$\left[\langle e^{-\beta_2 W}\rangle_{\mathrm{QA}}\right]_{\beta_1} = \sum_{\{J_{ij}\}} \frac{Z_T(\beta_2; \{J_{ij}\})Z_T(\beta_1; \{J_{ij}\})}{2^N (2\cosh\beta_1 J)^{N_B} (2\cosh\beta_2\Gamma_0)^N}. \tag{156}$$

Comparing Eqs. (155) and (156), we find the following relation between two different non-adiabatic processes,

$$\left[\langle e^{-\beta_1 W}\rangle_{\mathrm{QA}}\right]_{\beta_2} = \left[\langle e^{-\beta_2 W}\rangle_{\mathrm{QA}}\right]_{\beta_1} \left(\frac{\cosh\beta_1 J}{\cosh\beta_2 J}\right)^{N_B} \left(\frac{\cosh\beta_2\Gamma_0}{\cosh\beta_1\Gamma_0}\right)^N. \tag{157}$$

We describe the two different paths of NQA related by this equality in Fig. 20. Setting $\beta_2 = 0$ in Eq. (157), (implying $p = 1/2$, the symmetric distribution or the high-temperature limit), we find a simple equality on the performed work during NQA

$$\left[\langle e^{-\beta_1 W}\rangle_{\mathrm{QA}}\right]_0 = \frac{(\cosh\beta_1 J)^{N_B}}{(\cosh\beta_1\Gamma_0)^N}. \tag{158}$$

The symmetric distribution ($\beta_2 = 0$ on the left-hand side) makes it possible to reduce the right-hand side to the above trivial expression. It is remarkable that NQA, which involves very complex dynamics, satisfies such a simple identity irrespective of the speed of annealing T. If we apply the Jensen inequality to the above equality, we can obtain the lower bound for the performed work as

$$[\langle W\rangle_{\mathrm{QA}}]_0 \geq -\frac{N}{\beta}\log\left(\frac{(\cosh\beta J)^{\frac{N_B}{N}}}{\cosh\beta\Gamma_0}\right). \tag{159}$$

Here we generalize the inverse temperature to β from the specific choice β_1. This lower bound is loose, since the direct application of the Jensen inequality to JE for NQA yields, after the configurational average with the symmetric distribution,

$$[\langle W\rangle_{\mathrm{QA}}]_0 \geq \frac{1}{\beta}D(0|\beta) - \frac{N}{\beta}\log\left(\frac{(\cosh\beta J)^{\frac{N_B}{N}}}{\cosh\beta\Gamma_0}\right), \tag{160}$$

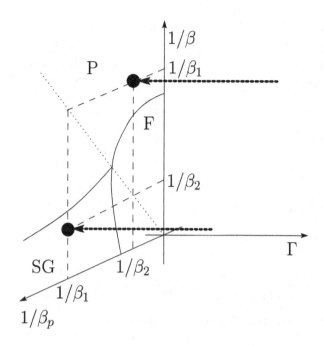

Fig. 20. Two different processes of NQA in Eq. (157). The left-hand side of Eq. (157) represents the annealing process ending at the upper-right black dot and the right-hand side terminates at the lower-left dot. Three phases expressed by the same symbols as in Fig. 12 are separated by solid curves and a vertical line. The dotted line expresses the NL $\beta_p = \beta$.

where $D(\beta|\beta')$ is the Kullback-Leibler divergence defined as

$$D(\beta|\beta') = \sum_{\{J_{ij}\}} \tilde{P}_\beta(\{J_{ij}\}) \log \frac{\tilde{P}_{\beta'}(\{J_{ij}\})}{\tilde{P}_\beta(\{J_{ij}\})}. \tag{161}$$

Here we defined the marginal distribution for the specific configuration $\{J_{ij}\}$ summed over all the possible gauge transformations,

$$\tilde{P}_\beta(\{J_{ij}\}) = \frac{1}{2^N} \sum_{\{\xi_i\}} \prod_{\langle ij \rangle} P(J_{ij}) = \frac{Z_T(\beta; \{J_{ij}\})}{2^N (2 \cosh \beta J)^{N_B}}. \tag{162}$$

Since the Kullback-Leibler divergence does not become non-negative, the work performed by the transverse field during a nonequilibrium process in in the symmetric distribution (i.e. the left-hand side of Eq. (160)) does not exceed the second quantity on the right-hand side of Eq. (160). This fact means Eq. (159) was looser.

4.4.3. *Exact relations involving inverse statistics*

Beyond the above results, we can perform further non-trivial analyses for the nonequilibrium process in the special conditions. Let us next take the configurational average of the inverse of the Jarzynski equality, Eq. (150), as

$$\left[\frac{1}{\langle e^{-\beta W}\rangle_{\mathrm{QA}}}\right]_{\beta_p} = \left[\frac{(2\cosh\beta\Gamma_0)^N}{Z_T(\beta;\{J_{ij}\})}\right]_{\beta_p}. \tag{163}$$

The application of the gauge transformation to the right-hand side yields

$$\left[\frac{1}{\langle e^{-\beta W}\rangle_{\mathrm{QA}}}\right]_{\beta_p} = \sum_{\{J_{ij}\}} \frac{\exp\left(\beta_p\sum_{\langle ij\rangle}J_{ij}\xi_i\xi_j\right)}{(2\cosh\beta_p J)^{N_B}}\frac{(2\cosh\beta\Gamma_0)^N}{Z_T(\beta;\{J_{ij}\})}. \tag{164}$$

By summing the right-hand side over all the possible configurations of $\{\xi_i\}$ and dividing the result by 2^N, we reach

$$\left[\frac{1}{\langle e^{-\beta W}\rangle_{\mathrm{QA}}}\right]_{\beta_p} = \sum_{\{J_{ij}\}} \frac{Z_T(\beta_p;\{J_{ij}\})}{2^N(2\cosh\beta_p J)^{N_B}}\frac{(2\cosh\beta\Gamma_0)^N}{Z_T(\beta;\{J_{ij}\})}. \tag{165}$$

If we set $\beta_p = \beta$ on the Nishimori line, this equation reduces to

$$\left[\frac{1}{\langle e^{-\beta W}\rangle_{\mathrm{QA}}}\right]_{\beta} = \frac{(\cosh\beta\Gamma_0)^N}{(\cosh\beta J)^{N_B}}. \tag{166}$$

Comparison of Eqs. (158) and (166) leads us to

$$[\langle e^{-\beta W}\rangle_{\mathrm{QA}}]_0 = \left(\left[\frac{1}{\langle e^{-\beta W}\rangle_{\mathrm{QA}}}\right]_{\beta}\right)^{-1}. \tag{167}$$

As shown in Fig. 21, two completely different processes are nontrivially related by the resultant relation: One toward the Nishimori line and the other for the symmetric distribution.

Let us further consider the inverse of Eq. (148) for the two-point correlation $O_T = \sigma_i^z\sigma_j^z$. We take the configurational average of both sides under the condition $\beta_p = \beta$ as

$$\left[\frac{1}{\langle \sigma_i^z\sigma_j^z e^{-\beta W}\rangle_{\mathrm{QA}}}\right]_{\beta} = \frac{(\cosh\beta\Gamma_0)^N}{(\cosh\beta J)^{N_B}}\left[\frac{1}{\langle \sigma_i^z\sigma_j^z\rangle_{\beta}}\right]_{\beta}. \tag{168}$$

The quantity on the right-hand side becomes unity by the gauge transformation as has been shown in the literatures.[7,22] We thus obtain a simple exact relation

$$\left[\frac{1}{\langle \sigma_i^z\sigma_j^z e^{-\beta W}\rangle_{\mathrm{QA}}}\right]_{\beta} = \frac{(\cosh\beta\Gamma_0)^N}{(\cosh\beta J)^{N_B}}, \tag{169}$$

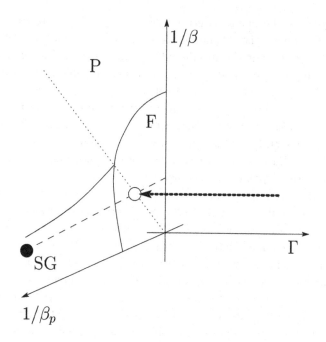

Fig. 21. Two different nonequilibrium processes in NQA through Eq. (167). We use the same symbols as in Fig. 12. The white circle denotes the target of the process on the right-hand side of Eq. (167), whereas the black dot is for the left-hand side.

which is another exact identity for processes of NQA.

The importance of the above equalities is still not clear. However it is true that we have little exact results on the nonequilibrium behavior of the spin glasses driven by quantum fluctuations as the transverse field. When we realize the quantum spin systems in experiments, the above results can play a roll as their indicators to verify their precisions and conditions. We should emphasize that the specialized tool to analyze spin glasses has facilitated the dynamical property in nonequilibrium process driven by quantum fluctuations. Beyond our analyses introduced here, we hope that more fascinating results on quantum computations would be obtained in the future with recourse to several techniques developed in statistical mechanics.

5. Summary

We looked over two topics lying between quantum information processing and statistical mechanics.

The first was the quantum error correction using the property of the topology. We prepare the redundant physical qubits in order to express the logical qubits we encode the specific information. Although we should deal with the quantum-many body systems, the used technique was proposed from the classical method in statistical mechanics. Statistical mechanics is available to facilitate to identify its precise location of the theoretical limitation to successfully infer the original state. In this direction with rich knowledge in statistical mechanics, we will be able to propose another ingenious way to give the quantum state more robustness and resilience as in the case of the depolarizing channel.

The second part was to improve the performance of quantum annealing away from the adiabatic-control case. The theoretical key was the Jarzynski equality. We suggest two ways to overcome the particular bottleneck in the adiabatic computation of quantum annealing. Both of the schemes would be needed for many-time repetition to produce the desired results. However we hope that the studies in this direction would give a novel technique beyond the ordinary limitations. The key point will be to enhance the possibility with the desired states. In the classical counterpart, recently several techniques are proposed with use of the population to reach the desired distribution, that is equilibrium state.[49-52] Such a technique as in the classical case enables us to generate the particular quantum state with higher probability than expected by use of skillful techniques.

The bridge between the quantum information processing and statistical mechanics continues to lead us to frontiers, where we encounter novel and surprising results beyond expected ones from knowledge we obtain only from each side. We are at a position to look at the birth of such a fascinating interdisciplinary field. Don't miss it.

References

1. A. Y. Kitaev, *Ann. Phys.* **303**, 2 (2003).
2. E. Dennis, A. Kitaev, A. Landahl, and J. Preskill, *J. Math. Phys.* **43**, 4452 (2002).
3. K. Binder and A. P. Young, *Rev. Mod. Phys.* **58**, 801 (1986).
4. A. P. Young (ed.), *Spin Glasses and Random Fields* (World Scientific, Singapore, 1997).
5. N. Kawashima and H. Rieger, in *Frustrated Spin Systems*, ed. T. H. Diep (World Scientific, Singapore, 2004).
6. M. Mézard, G. Parisi and M. A. Virasoro, *Spin Glass Theory and Beyond* (World Scientific, Singapore, 1987).
7. H. Nishimori, *Statistical Physics of Spin Glasses and Information Processing: An Introduction* (Oxford Univ. Press, Oxford, 2001).

8. P. W, Shor, *Phys. Rev. A* **52**, R2493 (1995).

9. A. B. Finnila, M. A. Gomez, C. Sebenik, S. Stenson and J. D. Doll, *Chem. Phys. Lett.* **219**, 343 (1994).

10. T. Kadowaki and H. Nishimori, *Phys. Rev. E* **58**, 5355 (1998).

11. T. Kadowaki, *Study of optimization problems by quantum annealing*, PhD thesis, Tokyo Institute of Technology, (1999); arXiv:0205020.

12. A. Das and B. K. Charkrabarti, *Quantum Annealing and Related Optimization Methods*, Lecture Notes in Physics Vol.679 Springer, Berlin, (2005).

13. G. E. Santoro and E. Tosatti, *J. Phys. A* **39**, R393 (2006).

14. A. Das and B. K. Chakrabarti, *Rev. Mod. Phys.* **80**, 1061 (2008).

15. S. Morita and H. Nishimori, *J. Math. Phys.* **49**, 125210 (2008).

16. M. Ohzeki and H. Nishimori, *J Comp. and Theor. Nanoscience* **8**, 963 (2011).

17. M. R. Garey and D. S. Johnson, *Computers and Intractability: A Guide to the Theory of NP-Completeness*, Freeman, San Francisco, (1979).

18. A. K. Hartmann and M. Weigt, *Phase Transitions in Combinatorial Optimization Problems: Basics, Algorithms and Statistical Mechanics* Wiley-VCH, Weinheim, (2005).

19. S. Kirkpatrick, S. D. Gelett and M. P. Vecchi, *Science* **220**, 671 (1983).

20. E. Aarts and J. Korst, *Simulated Annealing and Boltzmann Machines: A Stochastic Approach to Combinatorial Optimization and Neural Computing* (Wiley, New York, 1984).

21. C. W. Gardiner, *Handbook of Stochastic Methods* (Springer, 1985 2nd Edition).

22. H. Nishimori, *Prog. Theor. Phys.* **66**, 1169 (1981).

23. M. Ohzeki and H. Nishimori, *J. Phys. A: Math. Theor.* **42**, 332001 (2009).

24. H. Nishimori and K. Nemoto, *J. Phys. Soc. Jpn.* **71**, 1198 (2002).

25. J.-M. Maillard, K. Nemoto and H. Nishimori, *J. Phys. A* **36**, 9799 (2003).

26. M. Ohzeki, H. Nishimori and A. N. Berker, *Phys. Rev. E* **77**, 061116 (2008).

27. M. Ohzeki, *Phys. Rev. E* **79**, 021129 (2009).

28. H. A. Kramers and G. H. Wannier, *Phys. Rev.* **60**, 252 (1941).

29. F. Y. Wu and Y. K. Wang, *J. Math. Phys.* **17**, 439 (1976).

30. H. Nishimori, *J. Stat. Phys.* **126**, 977 (2007).

31. H. Bombin and M. A. Martin-Delgado, *Phys. Rev. A* **77**, 042322 (2008).

32. H. G. Katzgraber, H. Bombin and M.-A. Martin-Delgado, *Phys. Rev. Lett.* **103**, 090501 (2009).

33. M. Ohzeki, *Phys. Rev. E* **80**, 011141 (2009).

34. T. M. Stace, S. D. Barrett and A. C. Doherty, *Phys. Rev. Lett.* **102**, 200501 (2009).

35. T. M. Stace and S. D. Barrett, *Phys. Rev. A* **81**, 022317 (2010).

36. M. Ohzeki, arXiv:1202.2593.

37. H. Bombin, R. S. Andrist, M. Ohzeki, H. G. Katzgraber and M. A. Martin-Delgado, to appear in *Phys. Rev. X*.

38. T. Jörg, F. Krzakala, J. Kurchan and A. C. Maggs, *Phys. Rev. Lett.* **101**, 147204 (2008).

39. A. P. Young, S. Knysh and V. N. Smelyanskiy, *Phys. Rev. Lett.* **104**, 020502 (2010).

124

40. R. D. Somma, C. D. Batista and G. Ortiz, *Phys. Rev. Lett.* **99**, 030603 (2007).
41. M. Ohzeki, *Phys. Rev. Lett.* **105**, 050401 (2010).
42. C. Jarzynski, *Phys. Rev. Lett.* **78**, 2690 (1997).
43. C. Jarzynski, *Phys. Rev. E* **56**, 5018 (1997).
44. P. Wocjan, C. F. Chiang, D. Nagaj and A. Abeyesinghe, *Phys. Rev. A* **80**, 022340 (2009).
45. M. Ohzeki, H. Katsuda and H. Nishimori, *J. Phys. Soc. Jpn.* **80**, 084002 (2011).
46. H. Tasaki, arXiv:0009244.
47. M. Campisi, P. Talkner and P. Hänggi, *Phys. Rev. Lett.* **102**, 210401 (2009).
48. S. Morita, H. Nishimori and Y. Ozeki, *J. Phys. Soc. Jpn.* **75**, 014001 (2006).
49. R. M. Neal, *Statistics and Computing* **11**, 125 (2001).
50. Y. Iba, *Trans. Jpn. Soc. Artif. Intel.* **16**, 279 (2001).
51. K. Hukushima and Y. Iba, *AIP. Conf. Proc.* **690**, 200 (2003).
52. M. Ohzeki and H. Nishimori, *J. Phys. Soc. Jpn.* **79**, 084003 (2010).

SECOND LAW-LIKE INEQUALITIES WITH QUANTUM RELATIVE ENTROPY: AN INTRODUCTION

TAKAHIRO SAGAWA[1,2]

[1] *The Hakubi Center, Kyoto University,*
Yoshida-ushinomiya cho, Sakyo-ku, Kyoto, 606-8302, Japan
[2] *Yukawa Institute for Theoretical Physics, Kyoto University,*
Kitashirakawa-oiwake cho, Sakyo-ku, Kyoto, 606-8502, Japan
E-mail: sagawa@yukawa.kyoto-u.ac.jp

We review the basic properties of the quantum relative entropy for finite-dimensional Hilbert spaces. In particular, we focus on several inequalities that are related to the second law of thermodynamics, where the positivity and the monotonicity of the quantum relative entropy play key roles; these properties are directly applicable to derivations of the second law (e.g., the Clausius inequality). Moreover, the positivity is closely related to the quantum fluctuation theorem, while the monotonicity leads to a quantum version of the Hatano-Sasa inequality for nonequilibrium steady states. Based on the monotonicity, we also discuss the data processing inequality for the quantum mutual information, which has a similar mathematical structure to that of the second law. Moreover, we derive a generalized second law with quantum feedback control. In addition, we review a proof of the monotonicity in line with Petz.[100]

Keywords: Quantum relative entropy, Second law of thermodynamics, Quantum mutual information

1. Introduction

The quantum relative entropy[1,2] was introduced by Umegaki,[3] analogous to the classical relative entropy introduced by Kullback and Leibler.[4] In the early days,[5,6] a prominent result on the quantum relative entropy was a proof of the monotonicity,[7–9] based on the celebrated Lieb's theorem[10] and its application to the strong subadditivity of the von Neumann entropy.[11,12] Ever since, the quantum relative entropy has been widely applied to quantum information theory.[13–17] Moreover, the quantum relative entropy is useful to describe the mathematical structure of the second law of thermodynamics.[18] In fact, it has been shown that the quantum relative entropy is closely related to recent progresses in nonequilibrium statistical mechanics.

The fluctuation theorem is a remarkable result in modern nonequilibrium statistical mechanics, which characterizes a symmetry of fluctuations of the entropy production in thermodynamic systems.[19-53] From the fluctuation theorem, we can straightforwardly derive the second law of thermodynamics which states that the entropy production is nonnegative on average. It has been understood that the derivation of the second law based on the fluctuation theorem is closely related to the positivity of the relative entropy.[26-28,30,45] In fact, without invoking the fluctuation theorem, we can directly derive the second law of thermodynamics by using the positivity of the relative entropy.[30,54,55] If the microscopic dynamics of a thermodynamic system is explicitly described by quantum mechanics, the fluctuation theorem is referred to as the quantum fluctuation theorem,[30-53] which is one of the main topics of this article.

On the other hand, the monotonicity implies that the quantum relative entropy is non-increasing under any time evolution that occurs with unit probability in quantum open systems.[1,2,14,16,17,56,57] It has been known that the second law of thermodynamics can also be derived from the monotonicity of the quantum relative entropy.[58,59] Moreover, the monotonicity can be applied to describe transitions between nonequilibrium steady states (NESSs). In such situations, the monotonicity leads to a second law-like inequality, which we refer to as a quantum version of the Hatano-Sasa inequality.[58] In the classical regime, such an inequality can also be derived from a generalized fluctuation theorem called the Hatano-Sasa equality. It was first discussed for an overdamped Langevin system,[60] and has been applied to other situations and systems.[61-65] In this article, we discuss the quantum version of the Hatano-Sasa inequality, while the quantum Hatano-Sasa equality has not been fully understood.[52]

The monotonicity of the quantum relative entropy is also useful in quantum information theory.[14,16,17] In particular, the data processing inequality for the quantum mutual information can be directly derived from the monotonicity as is the case for the second law of thermodynamics. Moreover, based on the monotonicity, we can derive several important inequalities such as the Holevo bound, which identifies the upper bound of the accessible classical information that is encoded in quantum states.[66-68] On the other hand, a "dual" inequality of the Holevo bound is related to a quantitiy which we refer to as the QC-mutual information (or the Groenewold-Ozawa information).[69-71] It plays a key role in the formulation of a generalized second law of thermodynamics with quantum feedback control.[72-86]

This article aims to be an introduction to the quantum relative entropy for finite-dimensional Hilbert spaces, which is organized as follows.

In Sec. 2, we review the basic properties of quantum mechanics. In Sec. 2.1, we introduce quantum states and observables. In Sec. 2.2, we discuss dynamics of quantum systems, which is the main part of this section. In particular, we introduce the concept of completely-positive (CP) maps, and prove that any CP map has a Kraus representation in line with the proof by Choi.[87]

In Sec. 3, we introduce the von Neumann entropy and the quantum relative entropy. In particular, we discuss the monotonicity of the quantum relative entropy under completely-positive and trace-preserving (CPTP) maps in Sec. 3.3. The strong subadditivity of the von Neumann entropy is shown to be a straightforward consequence of the monotonicity of the quantum relative entropy.

In Sec. 4, we discuss the quantum mutual information and related quantities. In Sec. 4.1, we introduce the quantum mutual information, and prove the data processing inequality based on the monotonicity of the quantum relative entropy. In Sec. 4.2, we discuss the Holevo's χ-quantity, and prove the Holevo bound. In Sec. 4.3, we introduce the QC-mutual information and discuss its properties. In particular, we prove a "dual" inequality of the Holevo bound. We note that all of the three quantities reduce to the classical mutual information for classical cases.

In Sec. 5, we discuss some derivations of the second law of thermodynamics and its variants. In Sec. 5.1, we dicuss the relationship between thermodynamic entropy and the von Neumann entropy. In Sec. 5.2, we discuss a derivation based on the positivity of the quantum relative entropy. In Sec. 5.3, we discuss the quantum fluctuation theorem in a general setup, by introducing the stochastic entropy production. The quantum fluctuation theorem directly leads to the second law, which is equivalent to the derivation based on the positivity of the quantum relative entropy. In Sec. 5.4, we discuss a derivaton of the second law based on the monotonicity. We also discuss relaxation processes toward NESSs, and derive a quantum version of the Hatano-Sasa inequality for transitions between NESSs.

In Sec. 6, we discuss derivations of a generalized second law of thermodynamics with quantum feedback control, which involves the QC-mutual information. Our derivation is based on the positivity of the quantum relative entropy in Sec. 6.1, while on the monotonicity in Sec. 6.2.

In Sec. 7, as concluding remarks, we discuss the physical meanings and the validities of the foregoing derivations of the second law of thermodynamics in detail.

In Appendix A, we briefly summarize the basic concepts in the linear algebra. In Appendix B, we prove the monotonicity of the quantum relative entropy in line with Petz.[100]

2. Quantum States and Dynamics

First of all, we review basic concepts in quantum mechanics. We consider quantum systems described by finite-dimensional Hilbert spaces. Let $L(H, H')$ be the set of linear operators from Hilbert space H to H'. In particular, we write $L(H) := L(H, H)$ (see also Table 1). We note that the linear algebra with the bra-ket notation is briefly summarized in Appendix A.

Table 1. Symbols and their meanings.

Symbols	Meanings
$L(H, H')$	The set of linear operators from H to H'
$L(H)$	$L(H, H)$
$Q(H)$	$\{\rho \in L(H) : \rho \geq 0,\ \text{tr}[\rho] = 1\}$

2.1. *Quantum States and Observables*

In quantum mechanics, both quantum states and observables (physical quantities) can be described by operators on a Hilbert space. Let H be a Hilbert space that characterizes a quantum system. We define $Q(H) \subset L(H)$ such that any $\rho \in Q(H)$ satisfies

$$\rho \geq 0 \quad \text{and} \quad \text{tr}[\rho] = 1, \tag{1}$$

where $\rho \geq 0$ means that $\langle \psi | \rho | \psi \rangle \geq 0$ for any $|\psi\rangle \in H$ (or equivalently, ρ is positive),[a] and $\text{tr}[\rho]$ means the trace of ρ. We call $\rho \in Q(H)$ a density operator, which describes a quantum state. If the rank of ρ is one so that $\rho = |\psi\rangle\langle\psi|$ for $|\psi\rangle \in H$, ρ is called a pure state and $|\psi\rangle$ is called a state vector.[b] We note that any state vector satisfies $\langle\psi|\psi\rangle = 1$.

---

[a] In the present article, we follow the terminologies by Nielsen and Chuang[14] and say that a hermitian operator X is positive if $X \geq 0$ and positive definite if $X > 0$. See appendix A for details.

[b] In terms of the algebraic quantum theory using C^*-algebras, the definition of pure states depends on the choice of the algebra that is generated by observables. The above definition is valid only if the set of observables equals the set of all Hermitian operators in $L(H)$. In other words, we assumed that there is no superselection rule.

On the other hand, any Hermitian operator $X \in L(\boldsymbol{H})$ is called an observable, which describes a physical quantity such as a component of the spin of an atom. Any observable X is assumed to be measurable without any error in principle. Let $X := \sum_k x_k |\varphi_k\rangle\langle\varphi_k|$ be the spectrum decomposition, where $\{|\varphi_k\rangle\}$ is an orthonormal basis of $\boldsymbol{H}$. By the error-free measurement of observable X, the measurement outcome is given by one of x_k's. We note that k is also referred to as an outcome. The probability of obtaining x_k is given by

$$p(k) := \langle\varphi_k|\rho|\varphi_k\rangle, \tag{2}$$

where $\rho \in Q(\boldsymbol{H})$ is the density operator of the measured quantum state. Equality (2) is called the Born rule. The sum of the probabilities satisfies

$$\sum_k p(k) = \sum_k \langle\varphi_k|\rho|\varphi_k\rangle = \text{tr}[\rho] = 1. \tag{3}$$

We note that $\rho \geq 0$ and $\text{tr}[\rho] = 1$ respectively confirm $p(k) \geq 0$ and $\sum_k p(k) = 1$. The average of the outcomes is then given by

$$\langle X \rangle := \sum_k p(k)x_k = \text{tr}[X\rho], \tag{4}$$

which is a useful formula. If $\rho = |\psi\rangle\langle\psi|$ is a pure state, the Born rule (2) reduces to

$$p(k) = |\langle\varphi_k|\psi\rangle|^2, \tag{5}$$

and Eq. (4) to

$$\langle X \rangle = \langle\psi|X|\psi\rangle. \tag{6}$$

If there are two quantum systems A and B described by Hilbert spaces $\boldsymbol{H}_A$ and $\boldsymbol{H}_B$, their composite system AB is described by the tensor product $\boldsymbol{H}_A \otimes \boldsymbol{H}_B$. Let $\rho^{AB} \in Q(\boldsymbol{H}_A \otimes \boldsymbol{H}_B)$ be a density operator of the composite system. The partial states corresponding to A and B are respectively given by

$$\rho^A = \text{tr}_B[\rho^{AB}], \quad \rho^B = \text{tr}_A[\rho^{AB}]. \tag{7}$$

In fact, for any observable $X^A \in L(\boldsymbol{H}_A)$ and the identity $I^B \in L(\boldsymbol{H}_B)$, we have

$$\text{tr}_{AB}[(X^A \otimes I^B)\rho^{AB}] = \text{tr}_A[X^A \rho^A], \tag{8}$$

which is consistent with Eq. (7). If $\rho^{AB} = \rho^A \otimes \rho^B$ is satisfied, ρ^{AB} is called a product state. If a pure state is not a product state, it is called entangled.

We next show that any quantum state can be written as a pure state of an extended system including an auxiliary system. Let K be a set of indexes and $\rho = \sum_{k \in K} p_k |\psi_k\rangle\langle\psi_k| \in Q(\boldsymbol{H})$ be a state, where $|\psi_k\rangle$'s are not necessarily mutually-orthogonal. We introduce an auxiliary system R described by a Hilbert space $\boldsymbol{H}_R$ with an orthonormal basis $\{|r_k\rangle\}_{k \in K}$. By defining a pure state

$$|\Psi\rangle := \sum_{k \in K} \sqrt{p_k} |\psi_k\rangle |r_k\rangle \in \boldsymbol{H} \otimes \boldsymbol{H}_R, \tag{9}$$

we have

$$\rho = \mathrm{tr}_R[|\Psi\rangle\langle\Psi|]. \tag{10}$$

The state vector $|\Psi\rangle$ in Eq. (9) is called a purification of ρ. We note that a purification is not unique.

2.2. Quantum Dynamics

2.2.1. Unitary Evolution

The time evolution of an isolated quantum system is given by a unitary evolution. A density operator $\rho \in Q(\boldsymbol{H})$ evolves as

$$\rho \mapsto U\rho U^\dagger, \tag{11}$$

where $U \in L(\boldsymbol{H})$ is a unitary operator satisfying $U^\dagger U = UU^\dagger = I$. Any unitary evolution preserves the trace (i.e., $\mathrm{tr}[U\rho U^\dagger] = \mathrm{tr}[\rho]$) and the positivity of ρ. In the continuous-time picture, the time evolution of a density operator is given by the von Neumann equation:

$$\frac{d\rho(t)}{dt} = -\mathrm{i}[H, \rho(t)] := -\mathrm{i}(H\rho(t) - \rho(t)H), \tag{12}$$

where $H \in L(\boldsymbol{H})$ is a Hermitian operator called the Hamiltonian of the system. We set $\hbar = 1$ in this article. The unitary evolution from time 0 to t is described by unitary operator $U(t) = e^{-\mathrm{i}Ht}$. If we control the system by changing some external classical parameters such as a magnetic field, the Hamiltonian of the system can depend on time. In such a case, the von Neumann equation is given by

$$\frac{d\rho(t)}{dt} = -\mathrm{i}[H(t), \rho(t)], \tag{13}$$

which leads to

$$U(t) = \sum_{n=0}^{\infty} (-\mathrm{i})^n \int_0^t dt_1 \int_0^{t_1} dt_2 \cdots \int_0^{t_{n-1}} dt_n H(t_1) H(t_2) \cdots H(t_n)$$

$$=: \mathrm{T} \exp\left(-\mathrm{i} \int_0^t H(t') dt'\right),$$

(14)

where "T" represents the time-ordered product.

2.2.2. Completely Positive Maps

For an open quantum system that interacts with another quantum system, the time evolution is not given by a unitary evolution in general. Moreover, the input state (the initial state) and the output state (the final state) can be described by different Hilbert spaces with each other. The time evolution is generally described by a liner map $\mathcal{E} : L(\boldsymbol{H}) \to L(\boldsymbol{H'})$, where $\boldsymbol{H}$ and $\boldsymbol{H'}$ are the Hilbert spaces that describe the input and output systems, respectively.

For example, we consider a time evolution in which the input state is in $Q(\boldsymbol{H})$ and the output state is in $Q(\boldsymbol{H'})$. If another quantum system described by $\boldsymbol{H''}$ comes to interact with the input system and we access the total output state, then the output system becomes larger than the input one; the output system is described by $\boldsymbol{H'} = \boldsymbol{H} \otimes \boldsymbol{H''}$.

The general condition for $\mathcal{E}$ is given by the complete positivity. If $\mathcal{E}$ occurs with unit probability, it needs to be trace-preserving. We will discuss these two properties in detail.

Definition 2.1. $\mathcal{E}$ is called positive if $\mathcal{E}(X) \geq 0$ for any $X \in L(\boldsymbol{H})$ such that $X \geq 0$. Moreover, $\mathcal{E}$ is called completely positive (CP) if $\mathcal{E} \otimes \mathcal{I}_n$ is positive for any $n \in \mathbb{N}$, where $\mathcal{I}_n$ is the identity operator on $L(\mathbb{C}^n)$.

The positivity is enough to confirm the positivity of the output state in $Q(\boldsymbol{H'})$. On the other hand, the complete positivity confirms the positivity of the density operator of the total system including the environment. In fact, if a time evolution was not CP, one might observe a negative probability in the total system. We note that an important example of $\mathcal{E}$ that is positive but not CP is the transposition of operators with a matrix representation, which has been used for characterizing entanglements.[88,89]

The definition of the complete positivity can be rewritten as follows: $\mathcal{E}$ is CP if, for any $L(\boldsymbol{H})$-valued positive matrix $X_n := (X_{kl})_{1 \leq k,l \leq n}$ with $X_{kl} \in L(\boldsymbol{H})$, $L(\boldsymbol{H'})$-valued matrix $(\mathcal{E}(X_{kl}))_{1 \leq k,l \leq n}$ is also positive. The equivalence of the two definitions is confirmed as follows: X_n can be written as

$$X_n = \sum_{kl} X_{kl} \otimes |e_k\rangle\langle e_l| \in L(\boldsymbol{H} \otimes \mathbb{C}^n), \tag{15}$$

where $\{|e_k\rangle\}_{k=1}^n$ is an orthonormal basis of $\mathbb{C}^n$. We then have

$$(\mathcal{E} \otimes \mathcal{I}_n)(X_n) = \sum_{kl} \mathcal{E}(X_{kl}) \otimes |e_k\rangle\langle e_l| \in L(\boldsymbol{H}' \otimes \mathbb{C}^n). \tag{16}$$

We note that $\mathcal{E}$ is called n-positive if $\mathcal{E} \otimes \mathcal{I}_n$ is positive for a n.

Any time evolution in quantum systems needs to be CP. On the other hand, we introduce the second important property of $\mathcal{E}$:

Definition 2.2. We call $\mathcal{E} : L(\boldsymbol{H}) \to L(\boldsymbol{H}')$ trace-preserving (TP) if $\text{tr}[\mathcal{E}(X)] = \text{tr}[X]$ for any $X \in L(\boldsymbol{H})$.

This property confirms the conservation of the probability as $\text{tr}[\mathcal{E}(\rho)] = 1$ for $\rho \in Q(\boldsymbol{H})$. If $\mathcal{E}$ is both CP and TP, it is called CPTP (completely positive and trace-preserving). Any time evolution that occurs with unit probability needs to be CPTP.

The followings are simple examples.

Example 2.1. Any unitary evolution $\mathcal{E}(\rho) := U\rho U^\dagger$ is CPTP.

Example 2.2. Let $\rho \in L(\boldsymbol{H}_A \otimes \boldsymbol{H}_B)$. The partial trace $\mathcal{E}(\rho) := \text{tr}_B[\rho]$ is CPTP. In fact, $\mathcal{E}$ is CP, because, for any positive operator $\sigma \in L(\boldsymbol{H}_A \otimes \boldsymbol{H}_B \otimes \mathbb{C}^n)$, $(\mathcal{E} \otimes \mathcal{I}_n)(\sigma) = \text{tr}_B[\sigma]$ is also positive. $\mathcal{E}$ is TP, because $\text{tr}_A[\text{tr}_B[\rho]] = \text{tr}_{AB}[\rho]$.

Example 2.3. Let $\rho_A \in Q(\boldsymbol{H}_A)$. A map $\mathcal{E} : Q(\boldsymbol{H}) \to Q(\boldsymbol{H} \otimes \boldsymbol{H}_A)$ defined by $\mathcal{E}(\rho) := \rho \otimes \rho_A$ is CPTP.

We have the following example by combining the above three examples.

Example 2.4. Let $\rho_A \in Q(\boldsymbol{H}_A)$ be a state and $U \in Q(\boldsymbol{H} \otimes \boldsymbol{H}_A)$ be a unitary operator. A map $\mathcal{E} : Q(\boldsymbol{H}) \to Q(\boldsymbol{H})$ defined by $\mathcal{E}(\rho) := \text{tr}_A[U\rho \otimes \rho_A U^\dagger]$ is CPTP.

The above example is a typical description of the dynamics in open quantum systems. It will be shown in Sec. 2.2.4 that any CPTP map from $L(\boldsymbol{H})$ to $L(\boldsymbol{H})$ can be written in this form. The next example is a generalization of the above example to the cases that the input system and the output system are different.

Example 2.5. Let H and H' be Hilbert spaces corresponding to the input and the output systems, respectively. We introduce auxiliary systems described by H_A and H_B and assume that $H \otimes H_A \simeq H' \otimes H_B$. Let $\rho_A \in Q(H_A)$ be a state and $U \in L(H \otimes H_A) \simeq L(H' \otimes H_B)$ be a unitary operator. Then a linear map $\mathcal{E} : L(H) \to L(H')$ defined by $\mathcal{E}(\rho) := \mathrm{tr}_B[U\rho \otimes \rho_A U^\dagger]$ is CPTP.

We next consider quantum measurement processes.[14,90–95] Suppose that we perform a measurement on a quantum system and obtain outcome k with probability $p(k)$. Let $\rho \in Q(H)$ be the pre-measurement state and $\rho_k \in Q(H')$ be the post-measurement state with outcome k. We define a linear map $\mathcal{E}_k : Q(H) \to Q(H')$ such that

$$\rho_k = \frac{1}{p(k)}\mathcal{E}_k(\rho), \tag{17}$$

where the probability of obtaining outcome k is given by

$$p(k) = \mathrm{tr}[\mathcal{E}_k(\rho)]. \tag{18}$$

The map $\mathcal{E}_k$ needs to be CP, but it is not TP if $p(k) \neq 1$. In general, a time evolution that does not occur with unit probability does not need to be TP. We note that $\mathcal{E}_k$ needs to satisfy

$$\mathrm{tr}[\mathcal{E}_k(X)] \leq \mathrm{tr}[X] \tag{19}$$

for any positive $X \in L(H)$, since $p(k) \leq 1$ must hold for any $\rho \in Q(H)$. The ensemble average of ρ_k's over all outcomes is given by $\sum_k p(k)\rho_k = \sum_k \mathcal{E}_k(\rho) =: \mathcal{E}(\rho)$, where $\mathcal{E} := \sum_k \mathcal{E}_k$ is a CPTP map. We note that $\{\mathcal{E}_k\}$ is called an instrument, which characterizes the measurement process.[c]

We note that $\mathcal{E} : L(H) \to L(H')$ is called a unital map if it satisfies $\mathcal{E}(I) = I'$, where I and I' are the identities on H and H', respectively.

2.2.3. *Kraus Representation*

We next show that any CP map has a useful representation, which is called the Kraus representation.

Theorem 2.1 (Kraus representation). *A linear map* $\mathcal{E} : L(H) \to L(H')$ *is CP if and only if it can be written as*

$$\mathcal{E}(\rho) = \sum_k M_k \rho M_k^\dagger, \tag{20}$$

---

[c]Rigorously speaking, an instrument is a map from K' to $\sum_{k \in K'} \mathcal{E}_k$, where K' is an element of a σ-algebra over $K = \{k\}$.

where $\rho \in L(\boldsymbol{H})$, $M_k \in L(\boldsymbol{H}, \boldsymbol{H'})$, and the sum in the right-hand side (rhs) is taken over a finite number of k's. Equality (20) is called a Kraus representation, and M_k's are called Kraus operators.

Proof (Choi[87]).

(*Proof of* $\Rightarrow$) Suppose that $\mathcal{E}$ is CP. Let $\{|e_i\rangle\}_i$ be an orthonormal basis of $\boldsymbol{H}$. An operator $E := \sum_{ij} |e_i\rangle\langle e_j| \otimes |e_i\rangle\langle e_j|$ is positive because $\langle\psi|E|\psi\rangle = |\sum_i \langle e_i|\langle e_i|\psi\rangle|^2 \geq 0$ for any $|\psi\rangle \in \boldsymbol{H} \otimes \boldsymbol{H}$. Since $\mathcal{E}$ is CP, $\mathcal{E} \otimes \mathcal{I}$ is positive with $\mathcal{I}$ the identity on $L(\boldsymbol{H})$. Therefore, the following operator is positive:

$$(\mathcal{E} \otimes \mathcal{I})(E) = \sum_i \mathcal{E}(|e_i\rangle\langle e_j|) \otimes |e_i\rangle\langle e_j| \in L(\boldsymbol{H'} \otimes \boldsymbol{H}). \qquad (21)$$

Therefore, it has a spectrum decomposition of the form $(\mathcal{E} \otimes \mathcal{I})(E) = \sum_k |v_k\rangle\langle v_k|$ with $|v_k\rangle \in \boldsymbol{H'} \otimes \boldsymbol{H}$. We note that $|v_k\rangle$ can be written as $|v_k\rangle = \sum_i |x_k^i\rangle|e_i\rangle$ with $|x_k^i\rangle \in \boldsymbol{H'}$. We then obtain

$$(\mathcal{E} \otimes \mathcal{I})(E) = \sum_{kij} |x_k^i\rangle\langle x_k^j| \otimes |e_i\rangle\langle e_j|. \qquad (22)$$

In addition, $|x_k^i\rangle\langle x_k^j| = M_k |e_i\rangle\langle e_j| M_k^\dagger$ holds with $M_k := \sum_i |x_k^i\rangle\langle e_i| \in L(\boldsymbol{H}, \boldsymbol{H'})$. Therefore, we obtain

$$\mathcal{E}(|e_i\rangle\langle e_j|) = \sum_k M_k |e_i\rangle\langle e_j| M_k^\dagger \qquad (23)$$

for any (i, j), which implies Eq. (20).

(*Proof of* $\Leftarrow$) Suppose Eq. (20). Then $(\mathcal{E} \otimes \mathcal{I}_n)(\sigma) = \sum_k (M_k \otimes I_n)\sigma(M_k^\dagger \otimes I_n)$ holds for any $\sigma \in L(\boldsymbol{H} \otimes \mathbb{C}^n)$, where $\mathcal{I}_n$ and I_n are the identities on $L(\mathbb{C}^n)$ and $\mathbb{C}^n$, respectively. We then obtain, for any $|\psi\rangle \in \boldsymbol{H'} \otimes \mathbb{C}^n$,

$$\langle\psi|(\mathcal{E} \otimes \mathcal{I}_n)(\sigma)|\psi\rangle = \sum_k \langle\psi_k|\sigma|\psi_k\rangle, \qquad (24)$$

where $|\psi_k\rangle := M_k^\dagger \otimes I_n|\psi\rangle$. Therefore, $\langle\psi|(\mathcal{E} \otimes \mathcal{I}_n)(\sigma)|\psi\rangle$ is positive if σ is positive, which implies that $\mathcal{E} \otimes \mathcal{I}_n$ is positive and therefore $\mathcal{E}$ is CP. $\qquad\square$

Theorem 2.1 was first proved by Kraus[93] based on Stinespring's theorem.[92] The above proof is based on Choi's proof.[87]

We note that the Kraus representation is not unique. We also note that the above proof implies that the N-positivity is enough for $\mathcal{E}$ to have a Kraus representation with N the dimension of $\boldsymbol{H}$. The following theorem connects the condition of TP to the Kraus representation.

Theorem 2.2. *Let $\mathcal{E} : L(\boldsymbol{H}) \to L(\boldsymbol{H}')$ be a CP map. $\mathcal{E}$ is TP if and only if Kraus operators satisfy*

$$\sum_k M_k^\dagger M_k = I, \tag{25}$$

where I is the identity on $\boldsymbol{H}$.

Proof. Suppose that $\mathcal{E}$ is CPTP. We then have

$$\mathrm{tr}[\rho] = \mathrm{tr}[\mathcal{E}(\rho)] = \mathrm{tr}\left[\sum_k M_k^\dagger M_k \rho\right] \tag{26}$$

for any $\rho \in L(\boldsymbol{H})$, which implies Eq. (25). Conversely, if Eq. (25) is satisfied, $\mathrm{tr}[\rho] = \mathrm{tr}[\mathcal{E}(\rho)]$ holds. $\square$

We consider quantum measurements in terms of the Kraus representation. We first note that CP map $\mathcal{E}_k$ is written as

$$\mathcal{E}_k(\rho) = \sum_i M_{ki}\rho M_{ki}^\dagger. \tag{27}$$

Since $\mathcal{E} := \sum_k \mathcal{E}_k$ is assumed to be CPTP, we have

$$\sum_{ki} M_{ki}^\dagger M_{ki} = I. \tag{28}$$

The probability (18) of outcome k is then written as

$$p(k) = \mathrm{tr}\left[\sum_i M_{ki}\rho M_{ki}^\dagger\right] = \mathrm{tr}[E_k\rho], \tag{29}$$

where we defined

$$E_k := \sum_i M_{ki}^\dagger M_{ki}. \tag{30}$$

It is obvious that E_k's satisfy

$$E_k \geq 0 \tag{31}$$

and

$$\sum_k E_k = I. \tag{32}$$

Inequality (31) confirms the positivity of the probability (i.e., $p(k) \geq 0$), and Eq. (32) confirms that $\sum_k p(k) = 1$. Any $\{E_k\} \subset L(\boldsymbol{H})$ satisfying (31) and (32) is called the positive operator-valued measure (POVM).[d]

---

[d]Rigorously speaking, a POVM is a map from K' to $\sum_{k\in K'} E_k$, where K' is an element of a σ-algebra over $K = \{k\}$.

The projection measurement of an observable is a special case of the foregoing general formulation of quantum measurements. Let $X = \sum_k x_k P_k$ be an observable, where P_k's are projection operators. We assume that $x_k \neq x_{k'}$ for $k \neq k'$. If the Kraus representation of $\mathcal{E}_k$ is given by

$$\mathcal{E}_k(\rho) = P_k \rho P_k, \qquad (33)$$

then the measurement is called the projection measurement of X. In this case, the POVM consists of projection operators P_k, and the probability of outcome k is given by $p(k) = \text{tr}[P_k \rho]$, which is a slight generalization of the Born rule (2). We also refer to the projection measurement of X as that of $\{P_k\}$. In particular, if P_k is written as $P_k = |\psi_k\rangle\langle\psi_k|$ for any k, the projection measurement of X is referred to as that of orthonormal basis $\{|\psi_k\rangle\}$.

Example 2.6. We consider a simple model of a photodetection. Suppose that H is 2-dimensional and describes a two-level atom. Let $\{|0\rangle, |1\rangle\} \subset H$ be an orthonormal basis, where $|0\rangle$ and $|1\rangle$ respectively describe the ground state and the excited state. The atom emits a photon with probability p if it is in the excited state. We observe the photon number with unit efficiency, where the outcome is given by "0" or "1." In this case, the Kraus operators are given by

$$M_0 := |0\rangle\langle 0| + \sqrt{1-p}|1\rangle\langle 1|, \; M_1 := \sqrt{p}|0\rangle\langle 1|, \qquad (34)$$

which leads to the POVM that consists of

$$E_0 := |0\rangle\langle 0| + (1-p)|1\rangle\langle 1|, \; E_1 := p|1\rangle\langle 1|. \qquad (35)$$

If $p = 1$ holds, this measurement becomes the projection measurement of $X := x_0|0\rangle\langle 0| + x_1|1\rangle\langle 1|$, where we can define $x_0 := 0$ and $x_1 := 1$.

2.2.4. *Indirect Measurement Model*

We next show that any quantum measurement described by $\{\mathcal{E}_k\}$ with $\mathcal{E}_k : L(H) \to L(H)$ can be written by a simple model of an indirect measurement.

Theorem 2.3. *Let $\{\mathcal{E}_k\}$ be an instrument with $\mathcal{E}_k : L(H) \to L(H)$. There exist an auxiliary system R described by H_R, unitary operator $U \in L(H \otimes H_R)$, a reference state $|\phi_R\rangle \in H_R$, and projection operators $\{P_k\} \subset H_R$ satisfying $\sum_k P_k = I$, such that*

$$\mathcal{E}_k(\rho) = \text{tr}_R[(I \otimes P_k)U\rho \otimes |\phi_R\rangle\langle\phi_R|U^\dagger(I \otimes P_k)] \qquad (36)$$

for any ρ and k, where I is the identity on H.

Proof. Let $\{M_{ki}\}_i \subset L(\boldsymbol{H})$ be Kraus operators of $\mathcal{E}_k$, which is given by Eq. (27). We introduce an auxiliary system such that $\boldsymbol{H}_R$ has an orthonormal basis $\{|ki\rangle\}_{ki}$. We first show that, for any $|\phi_R\rangle \in \boldsymbol{H}_R$, there is a unitary operator $U \in L(\boldsymbol{H} \otimes \boldsymbol{H}_R)$ satisfying

$$U|\psi\rangle|\phi_R\rangle = \sum_{ki} M_{ki}|\psi\rangle|ki\rangle. \tag{37}$$

In fact, U conserves the inner product on $\{|\varphi\rangle|\phi_R\rangle; \ |\varphi\rangle \in \boldsymbol{H}\} \subset \boldsymbol{H} \otimes \boldsymbol{H}_R$, that is,

$$\left(\sum_{k'i'} |\langle\varphi|\langle k'i'|M_{k'i'}^\dagger\right)\left(\sum_{ki} M_{ki}|\psi\rangle|ki\rangle\right) = \sum_{ki}\langle\varphi|M_{ki}^\dagger M_{ki}|\psi\rangle = \langle\varphi|\psi\rangle \tag{38}$$

holds for any $|\psi\rangle, |\varphi\rangle \in \boldsymbol{H}$. We then have

$$U\rho \otimes |\phi_R\rangle\langle\phi_R|U^\dagger = \sum_{kik'i'} M_{ki}\rho M_{k'i'}^\dagger \otimes |ki\rangle\langle k'i'|. \tag{39}$$

By defining $P_k := \sum_i |ki\rangle\langle ki| \in L(\boldsymbol{H}_R)$, we have

$$(I \otimes P_k)U\rho \otimes |\phi_R\rangle\langle\phi_R|U^\dagger(I \otimes P_k) = \sum_{ii'} M_{ki}\rho M_{ki'}^\dagger \otimes |ki\rangle\langle ki'|, \tag{40}$$

and therefore

$$\text{tr}_R[(I \otimes P_k)U\rho \otimes |\phi_R\rangle\langle\phi_R|U^\dagger(I \otimes P_k)] = \sum_i M_{ki}\rho M_{ki}^\dagger = \mathcal{E}_k(\rho), \tag{41}$$

which implies (36). $\qquad\qquad\square$

Physically, R can be regarded as a probe system such as the local oscillator of a homodyne detection, and $|\phi_R\rangle$ as the initial state of the probe such as a coherent state. The measured system described by $\boldsymbol{H}$ interacts with the probe system by the unitary evolution. We next perform the projection measurement with $\{P_k\}$ on the probe, and obtain the information about ρ. The effect of this indirect measurement is characterized only by instrument $\{\mathcal{E}_k\}$. In the case of a CPTP map, Theorem 2.3 reduces to the following corollary.

Corollary 2.1. *Let $\mathcal{E} : L(\boldsymbol{H}) \to L(\boldsymbol{H})$ be a CPTP map. There exist an auxiliary system R described by $\boldsymbol{H}_R$, unitary operator $U \in L(\boldsymbol{H} \otimes \boldsymbol{H}_R)$, and a reference state $|\phi_R\rangle \in \boldsymbol{H}_R$ such that*

$$\mathcal{E}(\rho) = \text{tr}_R[U\rho \otimes |\phi_R\rangle\langle\phi_R|U^\dagger]. \tag{42}$$

The above corollary implies that any nonunitary evolution $\mathcal{E} : L(\boldsymbol{H}) \to L(\boldsymbol{H})$ can be modeled by a unitary evolution of an extended system.

We note that, in Theorem 2.3 and Corollary 2.1, the initial state of the total system described by $\boldsymbol{H} \otimes \boldsymbol{H}_R$ is a product state. In fact, if a CP map is reproduced by a single indirect measurement model for an arbitrary input state $\rho \in Q(\boldsymbol{H})$, then the initial state of R should be independent of ρ.

2.2.5. *Heisenberg Picture*

We briefly discuss the Schrödinger and the Heisenberg pictures of time evolutions. We define an inner product of $X, Y \in L(\boldsymbol{H})$ by

$$\langle X, Y \rangle_{\mathrm{HS}} := \mathrm{tr}[X^\dagger Y], \tag{43}$$

which is called the Hilbert-Schmidt inner product. We define the adjoint of a linear map $\mathcal{E} : L(\boldsymbol{H}) \to L(\boldsymbol{H}')$ in terms of the Hilbert-Schmidt inner product. A linear map $\mathcal{E}^\dagger : L(\boldsymbol{H}') \to L(\boldsymbol{H})$ is the adjoint of $\mathcal{E}$ if it satisfies

$$\langle \mathcal{E}^\dagger(X), Y \rangle_{\mathrm{HS}} = \langle X, \mathcal{E}(Y) \rangle_{\mathrm{HS}} \tag{44}$$

for any $X \in L(\boldsymbol{H}')$ and $Y \in L(\boldsymbol{H})$. If $\mathcal{E}$ is CP and its Kraus representation is given by Eq. (20), the adjoint of $\mathcal{E}$ is written as

$$\mathcal{E}^\dagger(X) = \sum_k M_k^\dagger X M_k, \tag{45}$$

which implies that $\mathcal{E}^\dagger$ is also CP from Theorem 2.1. We note that a CP map $\mathcal{E}$ is TP if and only if $\mathcal{E}^\dagger$ is unital, because $\mathcal{E}^\dagger(I) = \sum_k M_k^\dagger M_k$.

Let $\rho \in Q(\boldsymbol{H})$ be a state, $X \in L(\boldsymbol{H}')$ be an observable, and $\mathcal{E} : L(\boldsymbol{H}) \to L(\boldsymbol{H}')$ be a CP map. We then have

$$\mathrm{tr}[X \mathcal{E}(\rho)] = \mathrm{tr}[\mathcal{E}^\dagger(X) \rho], \tag{46}$$

where the left-hand side (lhs) is called the Schrödinger picture, while the rhs is called the Heisenberg picture.

3. Quantum Relative Entropy

We now introduce the quantum entropies and discuss their basic properties.

3.1. *Von Neumann Entropy*

We first introduce the von Neumann entropy.[90]

Definition 3.1. The von Neumann entropy of $\rho \in Q(\boldsymbol{H})$ is defined as

$$S(\rho) := -\mathrm{tr}[\rho \ln \rho]. \tag{47}$$

Remark 3.1. We note the relationship between the von Neumann entropy and the classical Shannon entropy.[96,97] Let $\{p(a)\}_{a \in A}$ be a probability distribution on a set A. We regard the distribution as a vector whose a-th entry is $p(a)$, which we denote as $\boldsymbol{p} := (p(a))_{a \in A}$. The Shannon entropy (or the Shannon information) of $\boldsymbol{p}$ is defined as

$$H(\boldsymbol{p}) := -\sum p(a) \ln p(a). \tag{48}$$

If the spectrum decomposition of ρ is given by $\rho = \sum p(a) |\varphi_a\rangle\langle\varphi_a|$ with an orthonormal basis $\{|\varphi_a\rangle\}_{a \in A}$, the von Neumann entropy of ρ reduces to the Shannon entropy of $\boldsymbol{p}$:

$$S(\rho) = H(\boldsymbol{p}). \tag{49}$$

The following theorems describe basic properties of the von Neumann entropy.

Theorem 3.1. *Let ρ_k's are density operators whose supports are mutually orthogonal. Then the von Neumann entropy of $\rho := \sum_k p_k \rho_k$ with $\sum p_k = 1$ satisfies*

$$S(\rho) = H(\boldsymbol{p}) + \sum_k p_k S(\rho_k), \tag{50}$$

where $H(\boldsymbol{p}) := -\sum_k p_k \ln p_k$.

Proof. Since the supports of ρ_k's are mutually orthogonal, we have

$$\begin{aligned}
S(\rho) &= -\sum_k \mathrm{tr}[p_k \rho_k \ln(p_k \rho_k)] \\
&= -\sum_k \mathrm{tr}[p_k \rho_k (\ln p_k + \ln \rho_k)] \\
&= -\sum_k p_k \ln p_k - \sum_k p_k \mathrm{tr}[\rho_k \ln \rho_k],
\end{aligned} \tag{51}$$

which implies Eq. (50). □

Theorem 3.2. *Let $|\Psi\rangle \in \boldsymbol{H}_A \otimes \boldsymbol{H}_B$ be a state vector of a composite system, whose partial states are given by $\rho^A := \mathrm{tr}_B[|\Psi\rangle\langle\Psi|]$ and $\rho^B := \mathrm{tr}_A[|\Psi\rangle\langle\Psi|]$. Then*

$$S(\rho^A) = S(\rho^B). \tag{52}$$

Proof. Let $|\Psi\rangle = \sum_k \sqrt{p_k}|\varphi_k\rangle|\psi_k\rangle$ be the Schmidt decomposition of $|\Psi\rangle$ with $\sum_k p_k = 1$. Then $\rho^A = \sum_k p_k|\varphi_k\rangle\langle\varphi_k|$ and $\rho^B = \sum_k p_k|\psi_k\rangle\langle\psi_k|$ hold, and therefore $S(\rho^A) = S(\rho^B) = -\sum_k p_k \ln p_k$. $\qquad\square$

3.2. *Quantum Relative Entropy and Its Positivity*

We now introduce the quantum relative entropy and prove its positivity.

Definition 3.2. Let $\rho, \sigma \in Q(H)$. The Quantum relative entropy of ρ to σ is defined as

$$S(\rho\|\sigma) := \text{tr}[\rho \ln \rho] - \text{tr}[\rho \ln \sigma]. \tag{53}$$

If there exists $|\psi\rangle \in H$ that satisfies $\sigma|\psi\rangle = 0$ and $\langle\psi|\rho|\psi\rangle \neq 0$, the quantum relative entropy is defined as $S(\rho\|\sigma) := +\infty$.

Remark 3.2. We note the relationship between the quantum relative entropy and the classical relative entropy (the Kullback-Leibler divergence).[4,97] Let $\boldsymbol{p} := (p(a))_{a \in A}$ and $\boldsymbol{q} := (q(a))_{a \in A}$ be probability distributions on a set A. The classical relative entropy of $\boldsymbol{p}$ to $\boldsymbol{q}$ is defined as

$$S(\boldsymbol{p}\|\boldsymbol{q}) := \sum_a p(a) \ln \frac{p(a)}{q(a)}. \tag{54}$$

If two density operators are given by $\rho = \sum_a p(a)|\varphi_a\rangle\langle\varphi_a|$ and $\sigma = \sum_a q(a)|\varphi_a\rangle\langle\varphi_a|$ with an orthonormal basis $\{|\varphi_a\rangle\}_{a \in A}$, the quantum relative entropy reduces to the classical one:

$$S(\rho\|\sigma) = S(\boldsymbol{p}\|\boldsymbol{q}). \tag{55}$$

We next prove the positivity of the quantum relative entropy, which plays a key role to derive the second law of thermodynamics in Secs. 5.2 and 5.3.

Theorem 3.3 (Positivity of the quantum relative entropy).

$$S(\rho\|\sigma) \geq 0, \tag{56}$$

where the equality is achieved if and only if $\rho = \sigma$. Inequality (56) is called the Klein inequality.

Proof. Let $\rho = \sum_a p(a)|\psi_a\rangle\langle\psi_a|$ and $\sigma = \sum_a q(a)|\varphi_a\rangle\langle\varphi_a|$, where $\{|\psi_a\rangle\}$ and $\{|\varphi_a\rangle\}$ are orthonormal bases. We first show that

$$S(\rho\|\sigma) \geq S(\rho\|\sigma'), \tag{57}$$

where $\sigma' := \sum_a q(a)|\psi_a\rangle\langle\psi_a|$. Inequality (57) is equivalent to $-\mathrm{tr}[\rho\ln\sigma] \geq -\mathrm{tr}[\rho\ln\sigma']$. We note that $-\mathrm{tr}[\rho\ln\sigma] = -\sum_a p(a)\langle\psi_a|\ln\sigma|\psi_a\rangle$ and $-\mathrm{tr}[\rho\ln\sigma'] = -\sum_a p(a)\ln\langle\psi_a|\sigma|\psi_a\rangle$. By applying the Jensen inequality to convex function $-\ln x$, we have

$$-\langle\psi_a|\ln\sigma|\psi_a\rangle = -\sum_b |\langle\psi_a|\varphi_b\rangle|^2 \ln q(b)$$

$$\geq -\ln\left(\sum_b |\langle\psi_a|\varphi_b\rangle|^2 q(b)\right) \qquad (58)$$

$$= -\ln\langle\psi_a|\sigma|\psi_a\rangle,$$

where we used that $\sum_b |\langle\psi_a|\varphi_b\rangle|^2 = 1$ holds for any a. Therefore, we obtain inequality (57). The equality in (57) is achieved if and only if $|\langle\psi_a|\varphi_{f(b)}\rangle|^2 = \delta_{ab}$, where $f(\cdot)$ is a bijection map and δ_{ab} is the Kronecker delta. By relabeling the indexes of $\{|\varphi_b\rangle\}$, we can choose $f(b) = b$ without loss of generality.

We next show

$$S(\rho\|\sigma') \geq 0, \qquad (59)$$

which is equivalent to the positivity of the classical relative entropy $\sum_a p(a)\ln(p(a)/q(a))$. By using inequality $\ln(x^{-1}) \geq 1 - x$ for $x > 0$, we obtain

$$\sum_a p(a)\ln\frac{p(a)}{q(a)} \geq \sum_a p(a)\left(1 - \frac{q(a)}{p(a)}\right) = 0, \qquad (60)$$

which implies inequality (59). The equality in (59) is achieved if and only if $p(a) = q(a)$ for any a.

By combining inequalities (57) and (59), we obtain inequality (56). The equality is achieved if and only if $|\langle\psi_a|\varphi_b\rangle|^2 = \delta_{ab}$ and $p(a) = q(a)$, which implies $\rho = \sigma$. □

By removing the assumptions that $\mathrm{tr}[\rho] = \sum_a p(a) = 1$ and $\mathrm{tr}[\sigma] = \sum_a q(a) = 1$ from the above proof, we can straightforwardly obtain a generalization of Eq. (56):

$$\mathrm{tr}[\rho(\ln\rho - \ln\sigma)] \geq \mathrm{tr}[\rho - \sigma] \qquad (61)$$

for any positive operators ρ and σ. Inequality (61) is also called the Klein inequality.

The subadditivity of the von Neumann entropy is a direct consequence of the positivity of the quantum relative entropy.

Theorem 3.4 (Subadditivity of von Neumann entropy). *Let $\rho \in Q(\boldsymbol{H}_A \otimes \boldsymbol{H}_B)$ be a density operator of a composite system, whose partial states are given by $\rho^A := \mathrm{tr}_B[\rho^{AB}] \in Q(\boldsymbol{H}_A)$ and $\rho^B := \mathrm{tr}_A[\rho^{AB}] \in Q(\boldsymbol{H}_B)$. Their von Neumann entropies satisfy*

$$S(\rho^{AB}) \leq S(\rho^A) + S(\rho^B), \tag{62}$$

where the equality is achieved if and only if $\rho^{AB} = \rho^A \otimes \rho^B$.

Proof. We have

$$S(\rho^A) + S(\rho^B) - S(\rho^{AB}) = S(\rho^{AB} \| \rho^A \otimes \rho^B) \geq 0, \tag{63}$$

where the right equality is achieved if and only if $\rho^{AB} = \rho^A \otimes \rho^B$. $\quad\square$

3.3. *Monotonicity of the Quantum Relative Entropy*

We next discuss that the quantum relative entropy is non-increasing under any CPTP map, which is called the monotonicity. The monotonicity can be applied to a derivation of the second law of thermodynamics and to a lot of theorems in quantum information theory.

Theorem 3.5 (Monotonicity of the quantum relative entropy). *Let $\rho, \sigma \in Q(\boldsymbol{H})$ be states and $\mathcal{E} : L(\boldsymbol{H}) \to L(\boldsymbol{H}')$ be a CPTP map. Then*

$$S(\mathcal{E}(\rho) \| \mathcal{E}(\sigma)) \leq S(\rho \| \sigma), \tag{64}$$

which is called the Uhlmann inequality.

Several proofs of the monotonicity have been known, which are not so simple. One of the proofs will be shown in Appendix B. The following corollary is a special case of the monotonicity.

Corollary 3.1. *Let $\rho^{AB}, \sigma^{AB} \in Q(\boldsymbol{H}_A \otimes \boldsymbol{H}_B)$ be density operators of a composite system, whose partial states are given by $\rho^A := \mathrm{tr}_B[\rho^{AB}]$ and $\sigma^A := \mathrm{tr}_B[\sigma^{AB}]$. Then*

$$S(\rho^A \| \sigma^A) \leq S(\rho^{AB} \| \sigma^{AB}). \tag{65}$$

Proof. Apply the monotonicity (64) to a CPTP map $\mathcal{E} : L(\boldsymbol{H}_A \otimes \boldsymbol{H}_B) \to L(\boldsymbol{H}_A)$ such that $\mathcal{E}(\rho) := \mathrm{tr}_B[\rho]$ for $\rho \in L(\boldsymbol{H}_A \otimes \boldsymbol{H}_B)$. $\quad\square$

We also have the following corollary, which implies that the von Neumann entropy is non-decreasing for a special class of CPTP maps.

Corollary 3.2. *Let* $\mathcal{E} : L(\boldsymbol{H}) \to L(\boldsymbol{H})$ *be a unital CPTP map satisfying* $\mathcal{E}(I) = I$, *where* I *is the identity on* $\boldsymbol{H}$. *The von Neumann entropy is then non-decreasing:*

$$S(\rho) \le S(\mathcal{E}(\rho)). \tag{66}$$

Proof. Let d be the dimension of $\boldsymbol{H}$. We than have

$$\begin{aligned} S(\rho) &= -S(\rho\|I/d) + \ln d \le -S(\mathcal{E}(\rho)\|\mathcal{E}(I/d)) + \ln d \\ &= -S(\mathcal{E}(\rho)\|I/d) + \ln d = S(\mathcal{E}(\rho)), \end{aligned} \tag{67}$$

where we used the monotonicity of the quantum relative entropy. $\square$

Let $\mathcal{E}(\rho) = \sum_k M_k \rho M_k^\dagger$ be the Kraus representation of $\mathcal{E}$. The condition of $\mathcal{E}(I) = I$ is satisfied if all of the Kraus operators are Hermitian such that $M_k = M_k^\dagger$. In particular, $\mathcal{E}(I) = I$ holds if M_k's are projection operators.

The strong subadditivity of the von Neumann entropy is easily obtained from the monotonicity of the quantum relative entropy.

Theorem 3.6 (Strong subadditivity of the von Neumann entropy). *Let* $\rho^{ABC} \in Q(\boldsymbol{H}_A \otimes \boldsymbol{H}_B \otimes \boldsymbol{H}_C)$, $\rho^{AB} := \mathrm{tr}_C[\rho^{ABC}]$, $\rho^{BC} := \mathrm{tr}_A[\rho^{ABC}]$, *and* $\rho^B := \mathrm{tr}_{AC}[\rho^{ABC}]$. *Then their von Neumann entropies satisfy*

$$S(\rho^{ABC}) + S(\rho^B) \le S(\rho^{AB}) + S(\rho^{BC}). \tag{68}$$

Proof. Let $\sigma^A := I_A/d_A$, where I_A is the identity on $\boldsymbol{H}_A$ and d_A is the dimension of $\boldsymbol{H}_A$. We then have

$$\begin{aligned} &[S(\rho^{AB}) + S(\rho^{BC})] - [S(\rho^{ABC}) + S(\rho^B)] \\ &= [S(\rho^{AB}) - S(\rho^{ABC})] - [S(\rho^B) - S(\rho^{BC})] \\ &= S(\rho^{ABC}\|\sigma^A \otimes \rho^{BC}) - S(\rho^{AB}\|\sigma^A \otimes \rho^B) \\ &\ge 0, \end{aligned} \tag{69}$$

where we used the monotonicity of the relative entropy for a CPTP map $\mathcal{E} : L(\boldsymbol{H}_A \otimes \boldsymbol{H}_B \otimes \boldsymbol{H}_C) \to L(\boldsymbol{H}_A \otimes \boldsymbol{H}_B)$ such that $\mathcal{E}(\rho) = \mathrm{tr}_C[\rho]$ for $\rho \in L(\boldsymbol{H}_A \otimes \boldsymbol{H}_B \otimes \boldsymbol{H}_C)$. $\square$

Historically, the strong subadditivity (68) of the von Neumann entropy was first proved based on the Lieb theorem.[10–12] Later, the monotonicity inequalities (64) and (65) for the quantum relative entropy were proved from the strong subadditivity (68).[7–9] On the other hand, Petz[98] proved monotonicity (65) without invoking the strong subadditivity. Nielsen and Petz[99] pedagogically discussed this proof. In a similar manner, Petz[100] showed a direct proof of monotonicity (64), which we will discuss in Appendix B.

Remark 3.3. In contrast to the quantum case, it is easy to prove the monotonicity of the classical relative entropy.[97] Let $p := (p(a))_{a \in A}$ and $q := (q(a))_{a \in A}$ be probability distributions on A. The classical counterpart of a quantum CPTP map is a Markov maps, which is given by transition probabilities $\{r(b|a)\}_{a \in A, b \in B}$ with $\sum_b r(b|a) = 1$ such that

$$p'(b) := \sum_a r(b|a)p(a), \quad q'(b) := \sum_a r(b|a)q(a). \tag{70}$$

We write $\mathcal{E}(p) := (p'(b))_{b \in B}$ and $\mathcal{E}(q) := (q'(b))_{b \in B}$. Our goal is to show

$$S(p\|q) \geq S(\mathcal{E}(p)\|\mathcal{E}(q)) \tag{71}$$

for the classical relative entropy. Let $p(a,b) := r(b|a)p(a)$ and $q(a,b) := r(b|a)q(a)$. We then have

$$\begin{aligned}
S(p\|q) &= \sum_{a,b} p(a,b) \ln \frac{p(a)}{q(a)} = \sum_{a,b} p(a,b) \ln \frac{p(a,b)}{q(a,b)} \\
&= S(\mathcal{E}(p)\|\mathcal{E}(q)) + \sum_{a,b} p(a,b) \ln \frac{p(a|b)}{q(a|b)},
\end{aligned} \tag{72}$$

where $p(a|b) := p(a,b)/p'(b)$ and $q(a|b) := q(a,b)/q'(b)$. By noting that $\sum_{a,b} p(a,b) \ln(p(a|b)/q(a|b)) \geq 0$ holds from the positivity of the classical relative entropy, we obtain inequality (71). We note that the strong subadditivity of the Shannon entropy straightforwardly follows from the monotonicity of the classical relative entropy.

4. Quantum Mutual Information and Related Quantities

In this section, we discuss the basic properties of the quantum mutual information and related quantities. In particular, we introduce two important quantities that are closely related to the quantum mutual information: the Holevo χ-quantity and the QC-mutual information (the Groenewold-Ozawa information). We discuss their information-theoretic meanings.

4.1. *Quantum Mutual Information*

We first introduce the quantum mutual information.

Definition 4.1. Let $\rho^{AB} \in Q(H_A \otimes H_B)$ be a quantum state and $\rho^A := \mathrm{tr}_B[\rho^{AB}]$ and $\rho^B := \mathrm{tr}_A[\rho^{AB}]$ be its partial states. The mutual information between two systems is then defined as

$$I^{A:B}(\rho^{AB}) := S(\rho^{AB} \| \rho^A \otimes \rho^B) = S(\rho^A) + S(\rho^B) - S(\rho^{AB}). \tag{73}$$

From the positivity of the quantum relative entropy,

$$I^{A:B}(\rho^{AB}) \geq 0 \tag{74}$$

holds, where the equality is achieved if and only if $\rho^{AB} = \rho^A \otimes \rho^B$.

Remark 4.1. We note the relationship between the quantum mutual information and the classical mutual information. Let $\{|\varphi_a\rangle\}_{a \in A}$ and $\{|\psi_b\rangle\}_{b \in B}$ be orthonormal bases of H_A and H_B, respectively. We define

$$\rho^{AB} := \sum_{ab} p(a,b) |\varphi_a\rangle\langle\varphi_a| \otimes |\psi_b\rangle\langle\psi_b|, \tag{75}$$

where $\{p(a,b)\}_{(a,b) \in A \times B}$ is a classical probability distribution on $A \times B$. Then the quantum mutual information reduces to

$$I^{A:B}(\rho^{AB}) = \sum_{a,b} p(a,b) \ln \frac{p(a,b)}{p(a)p(b)}, \tag{76}$$

where $p(a) := \sum_b p(a,b)$ and $p(b) := \sum_a p(a,b)$. The rhs of Eq. (76) is the classical mutual information between A and B.

We now discuss the data processing inequality, which is a straightforward consequence of monotonicity (64) of the quantum relative entropy.

Theorem 4.1 (Data processing inequality). *Let $\mathcal{E}_A : L(H_A) \to L(H_{A'})$ and $\mathcal{E}_B : L(H_B) \to L(H_{B'})$ be CPTP maps. The quantum mutual information is non-increasing by $\mathcal{E}_A \otimes \mathcal{E}_B$:*

$$I^{A':B'}((\mathcal{E}_A \otimes \mathcal{E}_B)(\rho^{AB})) \leq I^{A:B}(\rho^{AB}). \tag{77}$$

Proof. Noting that $(\mathcal{E}_A \otimes \mathcal{E}_B)(\rho^A \otimes \rho^B) = \mathcal{E}_A(\rho^A) \otimes \mathcal{E}_B(\rho^B)$ and

$$\mathrm{tr}_{B'}[(\mathcal{E}_A \otimes \mathcal{E}_B)(\rho^{AB})] = \mathcal{E}_A(\rho^A), \ \mathrm{tr}_{A'}[(\mathcal{E}_A \otimes \mathcal{E}_B)(\rho^{AB})] = \mathcal{E}_B(\rho^B), \tag{78}$$

inequality (77) follows from the monotonicity (64) of the quantum relative entropy. $\square$

The data processing inequality states that the quantum mutual information never increases by any CPTP map that is performed on each systems individually. The following corollary is a special case.

Corollary 4.1. *We consider three systems corresponding to H_A, H_B, and H_C. Then*

$$I^{A:B}(\rho^{AB}) \leq I^{A:BC}(\rho^{ABC}). \tag{79}$$

Proof. By taking $\mathcal{E}_{BC}(\rho^{BC}) := \mathrm{tr}_C[\rho^{BC}]$ and applying Theorem 4.1 to $\mathcal{I}_A \otimes \mathcal{E}_{BC}$, we obtain inequality (79). □

Remark 4.2. We note that inequality (79) can be written as

$$S(\rho^A) + S(\rho^B) - S(\rho^{AB}) \leq S(\rho^A) + S(\rho^{BC}) - S(\rho^{ABC}), \qquad (80)$$

which is equivalent to the strong subadditivity (68).

4.2. Holevo's χ-quantity

We next introduce the Holevo's χ-quantity (or just the χ-quantity) that is related to the accessible classical information encoded in quantum states.[66–68]

Definition 4.2. Let $a \in A$ be a classical probability variable, $p(a)$ be its distribution with $\sum_a p(a) = 1$, and $\rho_a \in Q(\boldsymbol{H}_S)$ be a quantum state labeled by a. The χ-quantity is defined as

$$\chi^{AS} := S(\rho) - \sum_a p(a)S(\rho_a), \qquad (81)$$

where $\rho := \sum_a p(a)\rho_a$.

We introduce an auxiliary system $\boldsymbol{H}_A$ with orthonormal basis $\{|\varphi_a\rangle\}_{a \in A}$ that can store the classical information about a. We define a density operator

$$\rho^{AS} := \sum_a p(a)|\varphi_a\rangle\langle\varphi_a| \otimes \rho_a. \qquad (82)$$

The χ-quantity is then given by the mutual information

$$\chi^{AS} = I^{A:S}(\rho^{AS}), \qquad (83)$$

which is a useful formula.

Remark 4.3. We note the relationship between χ-quantity and the classical mutual information. Let $\{|\psi_b\rangle\}_{b \in B}$ be an orthonormal basis of $\boldsymbol{H}_S$. We assume that ρ_a's can simultaneously be diagonalized as

$$\rho_a = \sum_b p(b|a)|\psi_b\rangle\langle\psi_b|, \qquad (84)$$

where $\sum_b p(b|a) = 1$ for any a. In this case, we can straightforwardly show that

$$\chi^{AS} = I^{A:B}, \qquad (85)$$

where $I^{A:B}$ is the classical mutual information between A and B for the joint distribution $p(a, b) := p(b|a)p(a)$.

The following theorems describe important properties of the χ-quantity and the von Neumann entropy.

Theorem 4.2 (Concavity of the von Neumann entropy). *The χ-quantity satisfies*

$$\chi^{AS} \geq 0, \tag{86}$$

or equivalently

$$S(\rho) \geq \sum_a p(a)S(\rho_a), \tag{87}$$

which is called the concavity of the von Neumann entropy. The equality is achieved if and only if $p(a) = 1$ for a single a.

Proof. The inequality is obvious from the positivity of the mutual information in Eq. (83). The equality is achieved if and only if ρ^{AS} is a product state, which implies that $p(a) = 1$ holds for a single a. □

Theorem 4.3. *The χ-quantity satisfies*

$$\chi^{AS} \leq H(\boldsymbol{p}), \tag{88}$$

or equivalently

$$S(\rho) \leq H(\boldsymbol{p}) + \sum_a p(a)S(\rho_a), \tag{89}$$

where $H(\boldsymbol{p}) := -\sum_a p(a)\ln p(a)$. The equality is achieved if the supports of ρ_a's are mutually orthogonal.

Proof. We introduce an auxiliary system $\boldsymbol{H}_{A'}$ with orthonormal basis $\{|\psi_a\rangle\}_{a \in A}$. We define a state

$$\sigma^{AA'} := \sum_a p(a)|\varphi_a\rangle\langle\varphi_a| \otimes |\psi_a\rangle\langle\psi_a| \in Q(\boldsymbol{H}_A \otimes \boldsymbol{H}_{A'}), \tag{90}$$

where the mutual information between A and A' is given by $I^{A:A'}(\sigma^{AA'}) = H(\boldsymbol{p})$. On the other hand, we define a CPTP map $\mathcal{E} : L(\boldsymbol{H}_{A'}) \to L(\boldsymbol{H}_S)$ such that

$$\mathcal{E}(|\psi_a\rangle\langle\psi_a|) = \rho_a \tag{91}$$

for any a. In fact, we can construct $\mathcal{E}$ satisfying Eq. (91) as follows. Let $\rho_a = \sum_i q_a(i)|ai\rangle\langle ai|$ be the spectrum decomposition of ρ_a. We define Kraus operators

$$M_{ai} := \sqrt{q_a(i)}|ai\rangle\langle\psi_a| \in L(\boldsymbol{H}_{A'}, \boldsymbol{H}_S), \tag{92}$$

which satisfies

$$\sum_{ai} M_{ai}^\dagger M_{ai} = \sum_{ai} q_a(i)|\psi_a\rangle\langle ai|ai\rangle\langle\psi_a| = \sum_a |\psi_a\rangle\langle\psi_a| = I^{A'}, \tag{93}$$

where $I^{A'}$ is the identity on $\boldsymbol{H}_{A'}$. By defining $\mathcal{E}$ with the Kraus operators (92), we have

$$\mathcal{E}(|\psi_a\rangle\langle\psi_a|) = \sum_i q_a(i)|ai\rangle\langle ai| = \rho_a, \tag{94}$$

which confirms Eq. (91).

By applying $\mathcal{E} \otimes \mathcal{I}_A$ to $\sigma^{AA'}$ with $\mathcal{I}_A$ the identity on $L(\boldsymbol{H}_A)$, we have

$$(\mathcal{E} \otimes \mathcal{I}_A)(\sigma^{AA'}) = \sum_a p(a)\rho_a \otimes |\varphi_a\rangle\langle\varphi_a| =: \rho^{AS}. \tag{95}$$

Therefore, from the data processing inequality (77), we obtain

$$H(\boldsymbol{p}) = I^{A:A'}(\sigma^{AA'}) \geq I^{A:S}(\rho^{AS}) = \chi^{AS}, \tag{96}$$

which implies inequality (89). If the supports of ρ_a's are mutually orthonormal, the equality in (89) is achieved because of Eq. (50). □

Theorem 4.3 implies that Eq. (50) is replaced by inequality (89) if the supports of ρ_a's are not mutually orthogonal. The following corollary is a direct consequence of Theorem 4.3.

Corollary 4.2. *We define $\rho := \sum_a p(a)|\phi_a\rangle\langle\phi_a|$ with $\sum_a p(a) = 1$, where $|\phi_a\rangle$'s are not necessarily mutually-orthogonal. We then have*

$$S(\rho) \leq H(\boldsymbol{p}), \tag{97}$$

where $H(\boldsymbol{p}) := -\sum_a p(a)\ln p(a)$.

Proof. Apply Theorem 4.3 to $\rho_a := |\phi_a\rangle\langle\phi_a|$. □

We next show the data processing inequality for the χ-quantity.

Theorem 4.4 (Data processing inequality). *We define*

$$\chi'^{AS'} := S(\mathcal{E}(\rho)) - \sum_a p(a)S(\mathcal{E}(\rho_a)), \tag{98}$$

where $\mathcal{E} : L(\boldsymbol{H}_S) \to L(\boldsymbol{H}_{S'})$ is a CPTP map. Then

$$\chi'^{AS'} \leq \chi^{AS}. \tag{99}$$

Proof. From Eq. (83) and inequality (77), we have $\chi'^{AS'} = I^{A:S'}((\mathcal{I}_A \otimes \mathcal{E})(\rho^{AS})) \leq I^{A:S}(\rho^{AS}) = \chi^{AS}$, where $\mathcal{I}_A$ is the identity on $L(\boldsymbol{H}_A)$. □

We next formulate and prove the Holevo bound, which determines the upper bound of the accessible classical information that is encoded in a quantum system. We consider that the classical information about $a \in A$ is encoded in a quantum state $\rho_a \in Q(\boldsymbol{H}_S)$. We extract the information about a by performing a quantum measurement on the quantum system.

Let $\{E_b\}_{b \in B}$ be a POVM. The probability of obtaining outcome b by a measurement with $\{E_b\}_{b \in B}$ on state ρ_a is given by

$$p(b|a) = \mathrm{tr}[E_b \rho_a]. \tag{100}$$

The joint probability of (a, b) is $p(a, b) = p(b|a)p(a)$, whose marginal distributions are $p(a) := \sum_b p(a, b)$ and $p(b) := \sum_a p(a, b)$. The mutual information $I^{A:B}$ between the two classical variables is then given by

$$I^{A:B} = \sum_{a,b} p(a, b) \ln \frac{p(a, b)}{p(a)p(b)}. \tag{101}$$

The Holevo bound states that the upper bound of the classical mutual information (101) is bounded by the χ-quantity.

Theorem 4.5 (Holevo bound).

$$I^{A:B} \leq \chi^{AS} \tag{102}$$

holds for any POVM $\{E_b\}_{b \in B}$.

Proof.[14] We introduce two auxiliary quantum systems described by Hilbert spaces $\boldsymbol{H}_A$ and $\boldsymbol{H}_B$. Their orthonormal bases are labeled by the corresponding classical variables as $\{|a\rangle\}_{a \in A} \subset \boldsymbol{H}_A$ and $\{|b\rangle\}_{b \in B} \subset \boldsymbol{H}_B$. We define $\rho^{AS} \in Q(\boldsymbol{H}_A \otimes \boldsymbol{H}_S)$ as

$$\rho^{AS} := \sum_a p(a)|a\rangle\langle a| \otimes \rho_a, \tag{103}$$

and define $\rho^{ASB} := \rho^{AS} \otimes |0\rangle\langle 0| \in Q(\boldsymbol{H}_A \otimes \boldsymbol{H}_S \otimes \boldsymbol{H}_B)$, where $|0\rangle \in \boldsymbol{H}_B$ is an initial reference state. It is easy to show that there is a CPTP map $\mathcal{E}_{SB}$ acting on $L(\boldsymbol{H}_S \otimes \boldsymbol{H}_B)$ such that

$$(\mathcal{I}_A \otimes \mathcal{E}_{SB})(\rho^{ABC}) = \sum_{a,b} p(a)|a\rangle\langle a| \otimes \sqrt{E_b}\rho_a\sqrt{E_b} \otimes |b\rangle\langle b| =: \rho'^{ASB}, \quad (104)$$

where $\mathcal{I}_A$ is the identity on $L(\boldsymbol{H}_A)$. We note that $\mathcal{E}_{SB}$ describes a measurement process corresponding to POVM $\{E_a\}$.

Let $\rho'^{AB} := \mathrm{tr}_S[\rho^{ASB}]$. We then obtain

$$\chi^{AS} = I^{A:S}(\rho^{AS}) = I^{A:SB}(\rho^{ASB}) \geq I^{A:SB}(\rho'^{ASB}) \geq I^{A:B}(\rho'^{AB}), \quad (105)$$

where we used Eq. (83) and the data processing inequalities (77) and (79). By noting that

$$\rho'^{AB} = \sum_{a,b} p(a,b)|a\rangle\langle a| \otimes |b\rangle\langle b|, \quad (106)$$

we obtain

$$I^{A:B}(\rho'^{AB}) = I^{A:B}, \quad (107)$$

where the rhs means the classical mutual information (101). Inequality (105) and Eq. (107) imply the Holevo bound (102). □

4.3. QC-mutual Information
(Groenewold-Ozawa Information)

We next introduce a quantity called the QC-mutual information that is also related to the accessible classical information encoded in quantum states.[69-71] We consider a quantum measurement described by POVM $\{E_b\}_{b \in B}$, where $B = \{b\}$ is the set of the outcomes. If the measured state is $\rho \in Q(\boldsymbol{H}_S)$, the probability of obtaining outcome b is given by $p(b) = \mathrm{tr}_S[E_b\rho]$. By defining

$$\rho_b := \frac{1}{p(b)}\sqrt{E_b}\rho\sqrt{E_b}, \quad (108)$$

we introduce the QC-mutual information as follows.

Definition 4.3. In the above setup, the QC-mutual information (the Groenewold-Ozawa information) is defined as

$$I_{QC}^{S:B} := S(\rho) - \sum_b p(b)S(\rho_b). \quad (109)$$

We note that the QC-mutual information (109) only depends on the measured state ρ and the POVM $\{E_b\}_{b \in B}$. We note that Groenewold[69] and Ozawa[70] originally discussed the case that Kraus operators $\{M_b\}_{b \in B} \subset \boldsymbol{H}_S$ satisfy $E_b := M_b^\dagger M_b$ for any $b \in B$.

Remark 4.4. We note the relationship between the QC-mutual information and the classical mutual information. Let $\{|\varphi_a\rangle\}_{a \in A}$ be an orthonormal basis of $\boldsymbol{H}_S$. We assume that ρ and E_b's are simultaneously diagonalized such that $\rho = \sum_a p(a)|\varphi_a\rangle\langle\varphi_a|$ and $E_b = \sum_a p(b|a)|\varphi_a\rangle\langle\varphi_a|$ for any $b \in B$. In this case, we have $\rho_b = \sum_a p(a|b)|\varphi_a\rangle\langle\varphi_a|$ with $p(a|b) := p(b|a)p(a)/\left(\sum_a p(b|a)p(a)\right)$. Therefore, we obtain

$$I_{\text{QC}}^{S:B} = I^{A:B}, \tag{110}$$

where $I^{A:B}$ is the classical mutual information between A and B for the joint distribution $p(a, b) := p(b|a)p(a)$.

We consider a quantum measurement with Kraus operators $\{M_b\}_{b \in B} \subset \boldsymbol{H}_S$ satisfying $E_b = M_b^\dagger M_b$. Let $\rho_b' := M_b \rho M_b^\dagger / p(b)$ and $\rho' := \sum_b p(b)\rho_b'$. The QC-mutual information can then be written as

$$I_{\text{QC}}^{S:B} = \chi^{SB} - \Delta S_{\text{meas}}, \tag{111}$$

where $\chi^{SB} := S(\rho') - \sum_b p(b)S(\rho_b')$ is the χ-quantity of the post-measurement states, and $\Delta S_{\text{meas}} := S(\rho') - S(\rho)$ is the change of the von Neumann entropy by the measurement.

On the other hand, the QC-mutual information $I_{\text{QC}}^{S:B}$ equals the χ-quantity of an auxiliary system. Let $\boldsymbol{H}_R$ be the Hilbert space of the auxiliary system R, and $|\Psi\rangle \in \boldsymbol{H}_S \otimes \boldsymbol{H}_R$ be a purification of ρ such that

$$\text{tr}_R[|\Psi\rangle\langle\Psi|] = \rho. \tag{112}$$

We define $\rho^R := \text{tr}_S[|\Psi\rangle\langle\Psi|]$ and

$$\rho_b^R := \text{tr}_S[(\sqrt{E_b} \otimes I^R)|\Psi\rangle\langle\Psi|(\sqrt{E_b} \otimes I^R)]/p(b), \tag{113}$$

where $I^R \in L(\boldsymbol{H}_R)$ is the identity. We note that $\rho^R = \sum_b p(b)\rho_b^R$. We then obtain the following theorem:

Theorem 4.6. *The QC-mutual information satisfies*

$$I_{\text{QC}}^{S:B} = \chi^{BR}, \tag{114}$$

where $\chi^{BR} := S(\rho^R) - \sum_b p(b)\rho_b^R$ *is the χ-quantity of* $\{\rho_b^R\}_{b \in B}$.

Proof. By noting that $\mathrm{tr}_R[(\sqrt{E_b} \otimes I^R)|\Psi\rangle\langle\Psi|(\sqrt{E_b} \otimes I^R)]/p(b) = \rho_b$, we have $S(\rho_b) = S(\rho_b^R)$ and $S(\rho) = S(\rho^R)$, which imply Eq. (114). $\qquad\square$

Therefore, the QC-mutual information satisfies the following inequality.

Corollary 4.3.

$$0 \leq I_{\mathrm{QC}}^{S:B} \leq H(\boldsymbol{p}), \qquad (115)$$

where $H(\boldsymbol{p}) := -\sum_b p(b) \ln p(b)$.

Proof. Apply inequalities (86) and (88) to χ^{BR}. $\qquad\square$

We next consider an information-theoretic meaning of the QC-mutual information. We assume that classical information about $a \in A$ is encoded in $\rho \in Q(\boldsymbol{H}_S)$ as

$$\rho = \sum_a q(a)\rho_a, \qquad (116)$$

where ρ_a's are density operators and $q(a)$'s satisfy $\sum_a q(a) = 1$. We then perform a measurement with POVM $\{E_b\}_{b\in B}$. The probability of obtaining b under the condition of a is given by

$$p(b|a) = \mathrm{tr}_S[E_b \rho_a]. \qquad (117)$$

The joint distribution is $p(a, b) := p(b|a)q(a)$. We note that the unconditional probability of obtaining b is given by $p(b) = \sum_a p(a, b) = \mathrm{tr}_S[E_a \rho]$, and the QC-mutual information is defined by Eq. (109) with $\rho_b := \sqrt{E_b}\rho\sqrt{E_b}/p(b)$. We then have the following theorem as a "dual" of the Holevo bound (Theorem 4.5).[71]

Theorem 4.7. *In the above setup, the classical mutual information $I^{A:B}$ between A and B is bounded by the QC-mutual information:*

$$I^{A:B} \leq I_{\mathrm{QC}}^{S:B}. \qquad (118)$$

Proof. Let $\rho_a = \sum_i q_a(i)|\psi_{ai}\rangle\langle\psi_{ai}|$ be the spectrum decomposition of ρ_a, where $\{|\psi_{ai}\rangle\}_i$ is an orthonormal basis of $\boldsymbol{H}_S$. We introduce an auxiliary system described by $\boldsymbol{H}_R$ with orthonormal basis $\{|r_{ai}\rangle\}_{ai}$, and define a purification of ρ:

$$|\Psi\rangle := \sum_{ai} \sqrt{q(a)q_a(i)}|\psi_{ai}\rangle|r_{ai}\rangle \in \boldsymbol{H}_S \otimes \boldsymbol{H}_R. \qquad (119)$$

We have $\rho = \mathrm{tr}_R[|\Psi\rangle\langle\Psi|]$ and $\rho_a = \mathrm{tr}_R[I^S \otimes P_a^R|\Psi\rangle\langle\Psi|]/q(a)$, where $P_a^R := \sum_i |r_{ai}\rangle\langle r_{ai}|$ and I^S is the identity on $\boldsymbol{H}_S$. On the other hand, we define $\rho^R := \mathrm{tr}_S[|\Psi\rangle\langle\Psi|]$ and $\rho_b^R := \mathrm{tr}_S[(\sqrt{E_b} \otimes I^R)|\Psi\rangle\langle\Psi|(\sqrt{E_b} \otimes I^R)]/p(b) = \mathrm{tr}_S[(E_b \otimes I^R)|\Psi\rangle\langle\Psi|]/p(b)$, where I^R is the identity on $\boldsymbol{H}_R$. By noting that

$$\mathrm{tr}_S\left[E_b\mathrm{tr}_R\left[I^S \otimes P_a^R|\Psi\rangle\langle\Psi|\right]\right] = \mathrm{tr}_R\left[P_a^R\mathrm{tr}_S\left[E_b \otimes I^R|\Psi\rangle\langle\Psi|\right]\right], \quad (120)$$

we obtain

$$\mathrm{tr}_S[E_b\rho_a]q(a) = \mathrm{tr}_R[P_a^R\rho_b^R]p(b), \quad (121)$$

and therefore

$$p(a,b) = \mathrm{tr}_R[P_a^R\rho_b^R]p(b). \quad (122)$$

Since $I_{\mathrm{QC}}^{S:B} = S(\rho^R) - \sum_b p(b)S(\rho_b^R)$ holds from Theorem 4.6, we obtain inequality (118) by applying the Holevo bound to $\{\rho_b^R\}_{b\in B}$. $\quad\square$

We note that, in the set up of the Holevo bound (Theorem 4.5), the encoding of the classical information is fixed and the measurement is arbitrary. In contrast, in the setup of Theorem 4.7, the encoding is arbitrary and the measurement is fixed. Inequality (118) determines the upper bound of the accessible classical information under the condition that the measurement is given by $\{E_b\}_{b\in B}$ and the ensemble average of the encoded states is given by ρ.

5. Second Law of Thermodynamics

We now derive the second law of thermodynamics in three manners. The first derivation is based on the positivity of the quantum relative entropy. The second derivation is based on the quantum fluctuation theorem, which is shown to be equivalent to the first derivation. The third one is based on the monotonicity of the quantum relative entropy.

5.1. *Thermodynamic Entropy and the von Neumann Entropy*

Before going to the main part of this section, we briefly discuss the relationship between the thermodynamic entropy and the von Neumann entropy. We consider a thermodynamic system that is in thermal equilibrium. Let H be the Hamiltonian of the system. The Helmholtz free energy is defined as

$$F := -\beta^{-1} \ln \mathrm{tr}[e^{-\beta H}], \quad (123)$$

where $\beta > 0$ is the inverse temperature of the system. The thermodynamic entropy S_{therm} satisfies

$$S_{\text{therm}} = \beta(\langle E \rangle - F), \tag{124}$$

where $\langle E \rangle$ is the average energy of the system. The thermodynamic relation (124) has been well established from the 19th century as a phenomenological thermodynamic relation for macroscopic systems. In terms of statistical mechanics, however, it is not so obvious to determine the microscopic expression of the thermodynamic entropy S_{therm} as will be discussed in Sec. 7.

As a special case, if we select the canonical distribution

$$\rho_{\text{can}} := e^{\beta(F-H)} \tag{125}$$

as a microscopic expression of the thermal equilibrium state, we can easily show that the thermodynamic entropy is given by the von Neumann entropy of the system:

Theorem 5.1. *The von Neumann entropy of ρ_{can} satisfies*

$$S(\rho_{\text{can}}) = \beta(\langle E \rangle_{\text{can}} - F), \tag{126}$$

where $\langle E \rangle_{\text{can}} := \text{tr}\,[H\rho_{\text{can}}]$ is the average energy in the canonical distribution.

The statistical-mechanical relation (126) is consistent with the thermodynamic relation (124) with the correspondence between $S(\rho_{\text{can}})$ and S_{therm}. Therefore, in the main part of this article, we will identify a canonical distribution to a thermal equilibrium state, and the von Neumann entropy to the thermodynamic entropy; these identifications have been widely used in statistical mechanics. Some subtle points on the validities of these identifications will be discussed in Sec. 7.

We can also easily calculate the quantum relative entropy of any state $\rho \in Q(\boldsymbol{H})$ to the canonical distribution ρ_{can} as

$$S(\rho \| \rho_{\text{can}}) = \beta(F - \langle E \rangle) - S(\rho), \tag{127}$$

where $\langle E \rangle := \text{tr}[H\rho]$ is the average energy of ρ. From the positivity of the quantum relative entropy, we have

$$S(\rho) \le \beta(F - \langle E \rangle), \tag{128}$$

where the equality is achieved only if $\rho = \rho_{\text{can}}$. Inequality (128) implies that the von Neumann entropy takes the maximum in the canonical distribution among the states that have the same amount of the energy.

5.2. From the Positivity of the Quantum Relative Entropy

We now derive the second law of thermodynamics based on the positivity of the quantum relative entropy and the unitarity of the time evolution of the system.[30,54] We first prove a very general but almost trivial equality and inequality, and next apply them to thermodynamic situations. Therefore, the nontrivial part of this subsection is in the applications to each examples.

We consider a unitary evolution of the system from state $\rho_i \in Q(H)$ to $\rho_f \in Q(H)$ such that $\rho_f = U\rho_i U^\dagger$. We also introduce a reference state $\rho_0 \in Q(H)$, which is different from ρ_i or ρ_f in general. Since $S(\rho_i) = S(\rho_f)$ holds due to the unitary evolution, we have the following theorem:

Theorem 5.2. *In the above setup,*

$$-\text{tr}[\rho_f \ln \rho_0] - S(\rho_i) = S(\rho_f \| \rho_0) \tag{129}$$

holds, which leads to

$$-\text{tr}[\rho_f \ln \rho_0] - S(\rho_i) \geq 0. \tag{130}$$

We note that we use the positivity of the relative entropy $S(\rho_f \| \rho_0)$ to derive inequality (130). While Eq. (129) and inequality (130) are obvious, they play key roles to derive the second law of thermodynamics as shown below. We discuss two typical situations in the following.

Example 5.1. We assume that the system is driven by a time-dependent Hamiltonian $H(t)$ from $t = 0$ to $t = \tau$, which gives the unitary operator as

$$U = \text{T} \exp\left(-\text{i} \int_0^\tau H(t)dt\right). \tag{131}$$

The free energy corresponding to the Hamiltonian at time t is given by

$$F(t) := -\beta^{-1}\text{tr}\left[e^{-\beta H(t)}\right], \tag{132}$$

where $\beta > 0$. We assume that the system is initially in the canonical distribution at inverse temperature β such that

$$\rho_i := e^{\beta(F(0)-H(0))}, \tag{133}$$

and define the reference state as

$$\rho_0 := e^{\beta(F(\tau)-H(\tau))}. \tag{134}$$

We stress that $\rho_f \neq \rho_0$ in general. In this setup, we have

$$\text{tr}\left[\rho_f \ln \rho_0\right] = \beta\left(F(\tau) - \text{tr}\left[H\rho_f\right]\right). \tag{135}$$

Therefore, Eq. (129) reduces to

$$\beta(\langle W \rangle - \Delta F) = S(\rho_{\mathrm{f}} \| \rho_0), \tag{136}$$

where

$$\langle W \rangle := \mathrm{tr}[\rho_{\mathrm{f}} H(\tau)] - \mathrm{tr}[\rho_{\mathrm{i}} H(0)] \tag{137}$$

is the energy difference of the system, and

$$\Delta F := F(\tau) - F(0) \tag{138}$$

is the free-energy difference corresponding to the initial and final Hamiltonians. We note that the energy difference $\langle W \rangle$ is regarded as the work performed on the system in this setup, because any heat bath is not attached to the system. Corresponding to inequality (130), we obtain the second law of thermodynamics

$$\langle W \rangle \geq \Delta F. \tag{139}$$

Example 5.2. We assume that the total system consists of the main system S and heat baths B_k ($k = 1, 2, \cdots$). The Hilbert spaces corresponding to S and B_k are respectively given by $\boldsymbol{H}_S$ and $\boldsymbol{H}_{B_k}$ so that $\boldsymbol{H} = \boldsymbol{H}_S \otimes_k \boldsymbol{H}_{B_k}$. Let $H^S \in L(\boldsymbol{H}_S)$ be the system's Hamiltonian, $H^{B_k} \in L(\boldsymbol{H}_{B_k})$ be the kth Bath's Hamiltonian, and $H^{SB_k} \in L(\boldsymbol{H}_S \otimes \boldsymbol{H}_{B_k})$ be the interaction Hamiltonian between S and B_k. We assume that H^S and H^{SB_k} are time-dependent, while H^{B_k} is time-independent. The total system then obeys a unitary evolution from $t = 0$ to $t = \tau$ corresponding to the total Hamiltonian

$$H(t) = H^S(t) + \sum_k (H^{SB_k}(t) + H^{B_k}), \tag{140}$$

where we omitted to write the tensor products with the identities on $\boldsymbol{H}_S$ and $\boldsymbol{H}_{B_k}$'s. Let $H^{\mathrm{int}}(t) := \sum_k H^{SB_k}(t)$. For simplicity, we assume that the interaction Hamiltonian satisfies $H^{\mathrm{int}}(0) = H^{\mathrm{int}}(\tau) = 0$.

Let $\rho_{\mathrm{can}}^{B_k}$ be the canonical distribution corresponding to H^{B_k} such that

$$\rho_{\mathrm{can}}^{B_k} := e^{\beta_k(F_k - H^{B_k})}, \tag{141}$$

where $\beta_k > 0$ is the inverse temperature of B_k, and

$$F_k := -\beta_k^{-1} \ln \mathrm{tr} \left[e^{-\beta H^{B_k}} \right]. \tag{142}$$

We assume that the initial state of the total system is given by a product state

$$\rho_{\mathrm{i}} := \rho_{\mathrm{i}}^S \otimes_k \rho_{\mathrm{can}}^{B_k}, \tag{143}$$

where ρ_i^S is an arbitrary initial state of the system. We note that Eq. (143) is consistent with assumption $H^{\text{int}}(0) = 0$. The final state is given by $\rho_f = U\rho_i U^\dagger$, where U is given by Eq. (131) with the total Hamiltonian (140). The final state of S is given by

$$\rho_f^S := \text{tr}_B[\rho_f], \tag{144}$$

where tr_B means the trace over all $\boldsymbol{H}_{B_k}$'s. We then define the reference state as

$$\rho_0 := \rho_f^S \otimes_k \rho_{\text{can}}^{B_k}. \tag{145}$$

In this setup, we can show that Eq. (129) is equivalent to

$$\Delta S - \sum_k \beta_k \langle Q_k \rangle = S(\rho_f \| \rho_0), \tag{146}$$

where

$$\Delta S := S(\rho_f^S) - S(\rho_i^S) \tag{147}$$

is the difference in the von Neumann entropy of the system, and

$$\langle Q_k \rangle := \text{tr}\left[H^{B_k} \rho_i\right] - \text{tr}\left[H^{B_k} \rho_f\right] \tag{148}$$

is regarded as the heat that is absorbed by S from bath B_k due to assumption $H^{\text{int}}(0) = H^{\text{int}}(\tau) = 0$. Corresponding to inequality (130), we obtain the Clausius inequality

$$\Delta S - \sum_k \beta_k \langle Q_k \rangle \geq 0. \tag{149}$$

We note that, in the conventional thermodynamics, the initial and final states of the system are assumed to be in thermal equilibrium. On the other hand, we assumed nothing on ρ_i^S and ρ_f^S above. Therefore, inequality (149) is regarded as a generalization of the conventional Clausius inequality to situations in which the initial and final states of the system are out of equilibrium.

In the following, we additionally assume that the initial state of the system is given by the canonical distribution at inverse temperature β such that

$$\rho_i^S = e^{\beta(F^S(0) - H^S(0))}, \tag{150}$$

where

$$F^S(t) := -\beta^{-1}\text{tr}\left[e^{-\beta H^S(t)}\right]. \tag{151}$$

This assumption is consistent with assumption $H^{\text{int}}(0) = 0$. By using notation $\rho_0^S := e^{\beta(F^S(\tau) - H^S(\tau))}$, we obtain

$$S(\rho_{\mathrm{f}}^S) \leq -\mathrm{tr}\left[\rho_{\mathrm{f}}^S \ln \rho_0^S\right] = \beta\left(F(\tau) - \mathrm{tr}\left[H^S \rho_{\mathrm{f}}^S\right]\right), \tag{152}$$

where we used the positivity of $S(\rho_{\mathrm{f}}^S \| \rho_0^S)$. Therefore, we obtain

$$\langle \Delta E^S \rangle - \Delta F^S \geq \Delta S, \tag{153}$$

where

$$\langle \Delta E^S \rangle := \mathrm{tr}[\rho_{\mathrm{f}}^S H^S(\tau)] - \mathrm{tr}[\rho_{\mathrm{i}}^S H^S(0)] \tag{154}$$

is the energy difference of the system. By combining inequalities (149) and (153), we obtain

$$\beta\left(\langle \Delta E^S \rangle - \Delta F^S\right) - \sum_k \beta_k \langle Q_k \rangle \geq 0. \tag{155}$$

For a special case in which there is a single heat bath at inverse temperature β, inequality (130) reduces to

$$\langle W \rangle \geq \Delta F^S, \tag{156}$$

where

$$\langle W \rangle := \langle E^S \rangle - \langle Q \rangle \tag{157}$$

is the work performed on the system.

The argument in this subsection is based on the positivity of the quantum relative entropy and the unitary evolution of the total system. We can replace the unitary evolution by a unital CPTP map $\mathcal{E}$ satisfying $\mathcal{E}(I) = I$. In this case, the von Neumann entropy of the total system is non-decreasing as $S(\rho_{\mathrm{i}}) \leq S(\rho_{\mathrm{f}})$, which has been shown in Corollary 3.2. Thus, Eq. (129) is replaced by an inequality

$$-\mathrm{tr}[\rho_{\mathrm{f}} \ln \rho_0] - S(\rho_{\mathrm{i}}) \geq S(\rho_{\mathrm{f}} \| \rho_0), \tag{158}$$

and therefore, inequality (130) remains unchanged. As a consequence, inequalities (139) and (149) still hold for such a CPTP map $\mathcal{E}$ acting on the total system.

5.3. *From the Quantum Fluctuation Theorem*

The quantum fluctuation theorem gives information about fluctuations of the entropy production.[30-53] We first introduce the stochastic entropy production and formulate the quantum fluctuation theorem in a very general setup. Let $\rho_i, \rho_0 \in Q(H)$ be density operators. They have spectrum decompositions $\rho_i = \sum_a p_i(a)|\psi_a\rangle\langle\psi_a|$ and $\rho_0 = \sum_b p_0(b)|\phi_b\rangle\langle\phi_b|$, where $\{|\psi_a\rangle\}$ and $\{|\phi_b\rangle\}$ are orthonormal basis of H. To formulate the quantum fluctuation theorem, the key concepts are the forward and backward processes that are described as follows.

Forward process. In the forward process, the initial state is given by ρ_i. We first perform the projection measurement on ρ_i with basis $\{|\psi_a\rangle\}$, and obtain outcome a with probability $p_i(a)$. By this measurement, the ensemble average of the post-measurement states equals ρ_i. We next perform a unitary operation with a time-dependent Hamiltonian $H(t)$ from $t = 0$ to τ. The unitary operator is given by Eq. (131). The density operator of the system then becomes $\rho_f = U\rho_i U^\dagger$. We next perform the projection measurement on ρ_f with basis $\{|\phi_b\rangle\}$, and obtain outcome b with probability $p_f(b) := \langle\phi_b|\rho_f|\phi_b\rangle$. The joint probability of (a, b) is given by

$$p(a, b) := p(b \leftarrow a)p_i(a), \tag{159}$$

where

$$p(b \leftarrow a) := |\langle\phi_b|U|\psi_a\rangle|^2 \tag{160}$$

is the transition probability. We note that $p_f(b) = \sum_a p(a, b)$.

Backward process. To formulate the backward process, we need to introduce the time-reversal operator Θ acting on H, which is an anti-unitary (i.e., inner-product preserving and anti-linear) operator satisfying $\Theta^2 = \Theta$ and $\Theta^\dagger = \Theta$. Here, an anti-linear operator satisfies that, for any $|\varphi_1\rangle, |\varphi_2\rangle \in H$ and $\alpha_1, \alpha_2 \in \mathbb{C}$,

$$\Theta(\alpha_1|\varphi_1\rangle + \alpha_2|\varphi_2\rangle) = \alpha_1^*\Theta|\varphi_1\rangle + \alpha_2^*\Theta|\varphi_2\rangle, \tag{161}$$

where α_i^* means the complex conjugate of α_i. Let $|\tilde{\psi}_a\rangle := \Theta|\psi_a\rangle$, $|\tilde{\phi}_b\rangle := \Theta|\phi_b\rangle$, and $\tilde{\rho}_0 := \Theta\rho_0\Theta = \sum_b p_0(b)|\tilde{\phi}_b\rangle\langle\tilde{\phi}_b|$. We note that $\{|\tilde{\psi}_a\rangle\}$ and $\{|\tilde{\phi}_b\rangle\}$ are orthonormal bases of H.

The protocol for the backward process is as follows. The initial state of the backward process is given by $\tilde{\rho}_0$. We first perform the projection measurement on $\tilde{\rho}_0$ with basis $\{|\tilde{\phi}_b\rangle\}$, and obtain outcome b with probability

$p_0(b)$. By this measurement, the ensemble average of the post-measurement states equals $\tilde{\rho}_0$. We introduce the time-reversal of the Hamiltonian as

$$\tilde{H}(t) := \Theta H(t)\Theta. \tag{162}$$

For example, if the Hamiltonian depends on magnetic field B as $H(t; B)$, its time-reversal is given by $\tilde{H}(t; B) = H(t; -B)$. We next perform a unitary operation from $t = 0$ to $t = \tau$ with the time-reversed control protocol of the time-reversed Hamiltonian. The corresponding unitary operator $\tilde{U}$ is given by

$$\tilde{U} := \mathrm{T} \exp\left(-\mathrm{i} \int_0^\tau \tilde{H}(\tau - t)dt\right). \tag{163}$$

We next perform the projection measurement on $\tilde{U}\tilde{\rho}_0\tilde{U}^\dagger$ with basis $\{|\tilde{\psi}_a\rangle\}$, and obtain outcome a with probability $\tilde{p}_{\mathrm{f}}(a) := \langle\tilde{\psi}_a|\tilde{U}\tilde{\rho}_0\tilde{U}^\dagger|\tilde{\psi}_a\rangle$. The joint probability of (b, a) in the backward process, denoted by $\tilde{p}(b, a)$, is then given by

$$\tilde{p}(b, a) = \tilde{p}(a \leftarrow b)p_0(b), \tag{164}$$

where

$$\tilde{p}(a \leftarrow b) = |\langle\tilde{\psi}_a|\tilde{U}|\tilde{\phi}_b\rangle|^2 \tag{165}$$

is the backward transition probability. We note that $\tilde{p}_{\mathrm{f}}(a) := \sum_b \tilde{p}(b, a)$.

We define the following quantity:

$$\sigma(a, b) := \ln\frac{p(a, b)}{\tilde{p}(b, a)}, \tag{166}$$

which is referred to as the stochastic entropy production in the forward process. The average of the entropy production is given by

$$\langle\sigma\rangle = \sum_{a,b} p(a, b) \ln\frac{p(a, b)}{\tilde{p}(b, a)}, \tag{167}$$

which is positive because of the positivity of the classical relative entropy:

$$\langle\sigma\rangle \geq 0. \tag{168}$$

We discuss the relationship between inequality (168) and inequality (130) in Theorem 5.2 in Sec. 5.2. The following theorem plays a key role.

Theorem 5.3. *The classical relative entropy (167) can be written as*

$$\langle\sigma\rangle = S(\rho_{\mathrm{f}}\|\rho_0), \tag{169}$$

where $S(\rho_f \| \rho_0)$ is the quantum relative entropy.

Proof. We first note that $\Theta \tilde{U} \Theta = U^\dagger$ holds, because $\Theta(i\tilde{H}(t))\Theta = -iH(t)$ holds. We then have the key observation that the unitary evolution has a time-reversal symmetry:

$$\tilde{p}(a \leftarrow b) = |\langle \tilde{\psi}_a | \tilde{U} | \tilde{\phi}_b \rangle|^2 = |\langle \psi_a | \Theta \tilde{U} \Theta | \phi_b \rangle|^2$$
$$= |\langle \psi_a | U^\dagger | \phi_b \rangle|^2 = |\langle \phi_b | U | \psi_a \rangle|^2 = p(b \leftarrow a). \tag{170}$$

Therefore, we obtain

$$\sigma(a, b) = \ln \frac{p_i(a)}{p_0(b)}, \tag{171}$$

which leads to

$$\langle \sigma \rangle = \sum_{a,b} p(a, b) \ln \frac{p_i(a)}{p_0(b)} = \sum_a p_i(a) \ln p_i(a) - \sum_b p_f(b) \ln p_0(b). \tag{172}$$

Obviously,

$$\sum_a p_i(a) \ln p_i(a) = -S(\rho_i) = -S(\rho_f). \tag{173}$$

We also obtain

$$\sum_b p_f(b) \ln p_0(b) = \sum_b \langle \phi_b | \rho_f | \phi_b \rangle \ln p_0(b)$$
$$= \sum_b \langle \phi_b | \rho_f \ln \rho_0 | \phi_b \rangle = \text{tr}[\rho_f \ln \rho_0]. \tag{174}$$

By combining Eqs. (173) and (174), we obtain Eq. (169). □

From Eqs. (129) and (166), we obtain

$$\langle \sigma \rangle = -\text{tr}[\rho_f \ln \rho_0] - S(\rho_i). \tag{175}$$

Therefore, inequality (168) is equivalent to inequality (130). Since inequality (130) leads to inequalities (139) and (149) as special cases, these inequalities can also be regarded as special cases of inequality (168), which we will discuss in detail later.

Equality (166) is regarded as a general expression of the quantum fluctuation theorem. The reason why Eq. (166) can be called a "theorem" rather than just a definition lies in the fact that σ equals to some important thermodynamic quantities for special cases, as we will show in Examples 5.3 and 5.4. Strictly speaking, Eq. (166) should be called a theorem only for such cases with the thermodynamic expressions of σ. In fact, we have already shown Eq. (175), which implies that the average of σ reduces to the lhs's of (136) and (149).

Before going to such special cases, we discuss the some properties of σ based on Eq. (166). To do so, we introduce the entropy production in the backward process as

$$\tilde{\sigma}(b,a) := \ln \frac{\tilde{p}(b,a)}{p(a,b)}. \tag{176}$$

In the backward process, $\Theta \rho_0 \Theta$ and $\Theta \rho_i \Theta$ respectively play the roles of ρ_i and ρ_0 in the forward process. Therefore, definition (176) in the backward process is consistent with the definition (167) in the forward process. We note that

$$\sigma(a,b) = -\tilde{\sigma}(b,a). \tag{177}$$

We introduce the probability distribution of σ as

$$p(\sigma = \Sigma) := \sum_{a,b} p(a,b)\delta(\Sigma, \sigma(a,b)), \tag{178}$$

where $\delta(\cdot, \cdot)$ is the Kronecker delta, and that of $\tilde{\sigma}$ as

$$\tilde{p}(\tilde{\sigma} = \Sigma) := \sum_{b,a} \tilde{p}(b,a)\delta(\Sigma, \tilde{\sigma}(b,a)). \tag{179}$$

We can show that

$$\frac{\tilde{p}(\tilde{\sigma} = -\Sigma)}{p(\sigma = \Sigma)} = e^{-\Sigma}, \tag{180}$$

because

$$\begin{aligned}
\tilde{p}(\tilde{\sigma} = -\Sigma) &= \sum_{a,b} \tilde{p}(b,a)\delta(-\Sigma, \tilde{\sigma}(b,a)) \\
&= \sum_{a,b} p(a,b)e^{\tilde{\sigma}(b,a)}\delta(-\Sigma, \tilde{\sigma}(b,a)) \\
&= e^{-\Sigma} \sum_{a,b} p(a,b)\delta(-\Sigma, \tilde{\sigma}(b,a)) \\
&= e^{-\Sigma} \sum_{a,b} p(a,b)\delta(\Sigma, \sigma(a,b)) \\
&= e^{-\Sigma} p(\sigma = \Sigma).
\end{aligned} \tag{181}$$

We also refer to Eq. (180) as the quantum fluctuation theorem. We can show that

$$\langle e^{-\sigma} \rangle = 1, \tag{182}$$

because

$$\langle e^{-\sigma} \rangle := \sum_{\Sigma} p(\sigma = \Sigma)e^{-\Sigma} = \sum_{\Sigma} \tilde{p}(\tilde{\sigma} = -\Sigma) = 1. \tag{183}$$

Equality (182) is called the integral fluctuation theorem or the quantum Jarzynski equality. By using the Jensen inequality for the exponential function (i.e., $e^{-\langle \sigma \rangle} \le \langle e^{-\sigma} \rangle$), we reproduce inequality (168) from Eq. (182). We note that the quantum fluctuation theorems (166), (180), and (182) were obtained by Kurchan[29] and Tasaki.[30]

Example 5.3. We consider the case of Example 5.1 in which ρ_i and ρ_0 are given by Eqs. (133) and (134), respectively. Let $H(0) = \sum_a E_a(0)|\psi_a\rangle\langle\psi_a|$ and $H(\tau) = \sum_b E_b(\tau)|\phi_b\rangle\langle\phi_b|$ be the spectrum decompositions of the initial and final Hamiltonians, which leads to $p_i(a) = e^{\beta(F(0)-E_a(0))}$ and $p_0(b) = e^{\beta(F(\tau)-E_b(\tau))}$. The stochastic entropy production (167) is then given by

$$\sigma(a,b) = \ln \frac{p_i(a)}{p_0(b)} = \beta(\Delta F - W(a,b)), \tag{184}$$

where $\Delta F := F(\tau) - F(0)$ and $W(a,b) := E_b(\tau) - E_a(0)$. We then obtain

$$\langle W \rangle := \sum_{a,b} p(a,b)W(a,b) = \text{tr}[H(\tau)\rho_f] - \text{tr}[H(0)\rho_i], \tag{185}$$

which is consistent with Eq. (137). In this case, the integral fluctuation theorem (182) reduces to

$$\langle e^{-\beta W} \rangle = e^{-\Delta F}, \tag{186}$$

which is called the quantum Jarzynski equality. Inequality (168) reduces to (139) in this situation.

Example 5.4. We next consider the case of Example 5.2 in which ρ_i and ρ_0 are given by Eqs. (143) and (145), respectively. Let $\rho_i^S := \sum_{a'} p_i^S(a')|\psi_{a'}^S\rangle\langle\psi_{a'}^S|$ and $\rho_f := \sum_{b'} p_f^S(b')|\phi_{b'}^S\rangle\langle\phi_{b'}^S|$ be the spectrum decompositions of the initial and final states of S. The spectrum decompositions of the Hamiltonians of the heat baths are given by $H^{B_k} = \sum_{a_k} E_{a_k}^{B_k}|\varphi_{a_k}\rangle\langle\varphi_{a_k}|$. We then have

$$p_i(a) = p_i^S(a') \prod_k \exp(\beta_k(F_k - E_{a_k}^{B_k})),$$

$$p_0(b) = p_f^S(b') \prod_k \exp(\beta_k(F_k - E_{b_k}^{B_k})), \tag{187}$$

where $a = (a', \{a_k\})$ and $b = (b', \{b_k\})$. We note that $|\psi_a\rangle = |\psi_{a'}^S\rangle \otimes_k |\varphi_{a_k}\rangle$ and $|\phi_b\rangle = |\psi_{b'}^S\rangle \otimes_k |\varphi_{b_k}\rangle$. Therefore, the stochastic entropy production is given by

$$\sigma(a, b) = s_f(b') - s_i(a') - \sum_k \beta_k Q_k(a_k, b_k), \tag{188}$$

where

$$s_i(a') := -\ln p_i^S(a'), \quad s_f(b') := -\ln p_f^S(b') \tag{189}$$

are called the stochastic entropies of S, and

$$Q_k(a_k, b_k) := E_{a_k}^{B_k} - E_{b_k}^{B_k} \tag{190}$$

is the heat absorbed by S from B_k. We note that

$$\langle s_i \rangle := \sum_{a,b} p(a,b) s_i(a') = -\sum_{a'} p_i^S(a') \ln p_i^S(a') = S(\rho_i^S),$$

$$\langle s_f \rangle := \sum_{a,b} p(a,b) s_f(b') = -\sum_{b'} p_f^S(b') \ln p_f^S(b') = S(\rho_f^S), \tag{191}$$

and

$$\langle Q_k \rangle := \sum_{a,b} p(a,b) Q_k(a_k, b_k) = \mathrm{tr}\left[H^{B_k} \rho_i\right] - \mathrm{tr}\left[H^{B_k} \rho_f\right], \tag{192}$$

which is consistent with Eq. (148). Therefore, we obtain

$$\langle \sigma \rangle = \Delta S - \sum_k \beta_k \langle Q_k \rangle, \tag{193}$$

where $\Delta S := S(\rho_f^S) - S(\rho_i^S)$. Inequality (168) reduces to the Clausius inequality (149) in this situation.

We note that the quantum fluctuation theorem has been generalized to nonunitary processes including quantum measurements.[48]

5.4. From the Monotonicity of the Quantum Relative Entropy

We next apply the monotonicity of the quantum relative entropy to a derivation of the second law of thermodynamics. While the obtained inequalities are mathematically not equivalent to the inequalities obtained in the previous two sections, their physical meanings are the same for special cases. The inequalities in this subsection can also be applied to transitions between nonequilibrium steady states, which leads to a quantum version of the Hatano-Sasa inequality.[58]

5.4.1. *Time-independent Control*

We first consider relaxation processes to steady states, in which the external parameters that we can control do not depend on time. The following theorem plays a key role.

Theorem 5.4. *Let $\mathcal{E} : L(\boldsymbol{H}) \to L(\boldsymbol{H})$ be a CPTP map and $\rho_i \in Q(\boldsymbol{H})$ be an initial state. We assume that $\mathcal{E}$ has a unique steady state ρ_{ss} satisfying $\mathcal{E}(\rho_{ss}) = \rho_{ss}$. Then*

$$\Delta S \geq -\sigma_{ex}^{B}, \tag{194}$$

where $\Delta S := S(\mathcal{E}(\rho_i)) - S(\rho_i)$ and

$$\sigma_{ex}^{B} := \mathrm{tr}[\mathcal{E}(\rho_i) \ln \rho_{ss}] - \mathrm{tr}[\rho_i \ln \rho_{ss}]. \tag{195}$$

Proof. Inequality (194) is obvious from the monotonicity of the relative entropy: $S(\rho_i \| \rho_{ss}) \geq S(\mathcal{E}(\rho_i) \| \mathcal{E}(\rho_{ss}))$ with $\mathcal{E}(\rho_{ss}) = \rho_{ss}$. $\square$

We apply inequality (194) to relaxation processes to equilibrium states.

Example 5.5. Suppose that $\rho_{ss} = I/d$ holds, where I is the identity on $\boldsymbol{H}$ and d is the dimension of $\boldsymbol{H}$. In this case, $\sigma_{ex}^{B} = 0$ holds, and therefore inequality (194) reduces to

$$\Delta S \geq 0, \tag{196}$$

which is equivalent to inequality (66).

Physically, the condition of $\mathcal{E}(I) = I$ implies that the steady state is the microcanonical distribution. In fact, the microcanonical distribution is given by $\rho_{microcan} := I/d$, where $\boldsymbol{H}$ is taken as the set of state vectors in a microcanonical energy shell. Thus, inequality (196) is regarded as the law of entropy increase for adiabatic processes, in which the system does not exchange the energy with the environment so that the steady state is expected to be the microcanonical distribution.

Example 5.6. Suppose that $\rho_{ss} = e^{\beta(F-H)}$ holds, where H is the Hamiltonian and $F := -\beta^{-1} \ln \mathrm{tr}[e^{-\beta H}]$ is the corresponding free energy. In this case, σ_{ex}^{B} is given by

$$\sigma_{ex}^{B} = -\beta(\mathrm{tr}[H\mathcal{E}(\rho_i)] - \mathrm{tr}[H\rho_i]) = -\beta\langle Q \rangle, \tag{197}$$

where Q is the heat absorbed by the system. Therefore, inequality (194) reduces to[58,59]

$$\Delta S \geq \beta Q, \tag{198}$$

which is the Clausius inequality with a single heat bath.

168

We note that inequality (194) can be applied to situations in which ρ_{ss} describes a nonequilibrium steady state (NESS). However, in such a case, the physical meaning of σ_{ex}^B is not so clear. For the case of a classical overdamped Langevin system, Hatano and Sasa[60] showed that σ_{ex}^B can be regarded as an "excess heat," which is obtained by subtracting a "housekeeping heat" from the total heat. The housekeeping heat means the heat current in a NESS, which vanishes for the cases of an equilibrium steady state.

5.4.2. *Time-dependent Control*

We next consider situations in which we drive a system by changing external parameters. We assume that we change the values of the parameters $N - 1$ times during the entire time evolution. In such a case, the total time evolution $\mathcal{E}$ can be written as

$$\mathcal{E} = \mathcal{E}_N \circ \cdots \circ \mathcal{E}_2 \circ \mathcal{E}_1, \tag{199}$$

where $\mathcal{E}_n : L(\boldsymbol{H}) \to L(\boldsymbol{H})$ ($n = 1, 2, \cdots, N$) describes the time evolution with the nth values of the external parameters. We assume that each $\mathcal{E}_n$ has a unique steady state $\rho_{ss,n}$. We write $\rho_n := \mathcal{E}_n(\rho_{n-1})$ and $\rho_0 := \rho_i$, where $\rho_i \in Q(\boldsymbol{H})$ is the initial state. From inequality (194), we have a set of inequalities

$$S(\rho_{n+1}) - S(\rho_n) \geq -\left(\text{tr}[\rho_n \ln \rho_{ss,n}] - \text{tr}[\rho_{n-1} \ln \rho_{ss,n}]\right) \ (1 \leq n \leq N). \tag{200}$$

By summing up them, we have the following theorem.

Theorem 5.5 (Quantum Hatano-Sasa inequality).

$$\Delta S \geq -\sigma_{ex}^B, \tag{201}$$

where

$$\sigma_{ex}^B := \sum_{n=1}^N \left(\text{tr}[\rho_n \ln \rho_{ss,n}] - \text{tr}[\rho_{n-1} \ln \rho_{ss,n}]\right). \tag{202}$$

If $\rho_{ss,n}$'s are out of equilibrium, inequality (201) is regarded as a quantum version of the Hatano-Sasa inequality, which was obtained by Yukawa.[58] We note that Eq. (202) can be rewritten as

$$\sigma_{ex}^B = \text{tr}[\rho_N \ln \rho_{ss,N}] - \text{tr}[\rho_0 \ln \rho_{ss,1}] - \sum_{n=1}^{N-1} \left(\text{tr}[\rho_n \ln \rho_{ss,n+1}] - \text{tr}[\rho_n \ln \rho_{ss,n}]\right). \tag{203}$$

If the initial and final states are the steady states such that $\rho_0 = \rho_{ss,1}$ and $\rho_N = \rho_{ss,N}$, we have $\sigma_{ex}^B = -\Delta S - \sum_{n=1}^{N-1}(\text{tr}[\rho_n \ln \rho_{ss,n+1}] - \text{tr}[\rho_n \ln \rho_{ss,n}])$. In this case, inequality (194) reduces to

$$\sum_{n=1}^{N-1}(\text{tr}[\rho_n \ln \rho_{ss,n+1}] - \text{tr}[\rho_n \ln \rho_{ss,n}]) \leq 0. \qquad (204)$$

If $\rho_{ss,n}$'s are equilibrium states, we again have the second law of thermodynamics as follows.

Example 5.7. Suppose that $\rho_{ss,n} = e^{\beta(F_n - H_n)}$ holds, where H_n is the Hamiltonian during $\mathcal{E}_n$ and $F_n := -\beta^{-1}\ln \text{tr}[e^{-\beta H_n}]$ is the corresponding free energy. In this case, σ_{ex}^B is given by

$$\sigma_{ex}^B = -\beta \sum_{n=1}^{N}(\text{tr}[H_n \rho_n] - \text{tr}[H_n \rho_{n-1}]) =: -\beta \langle Q \rangle, \qquad (205)$$

where $\langle Q \rangle$ is the heat absorbed by the system. Therefore, inequality (194) again reduces to

$$\Delta S \geq \beta \langle Q \rangle. \qquad (206)$$

We note that inequality (206) is not equivalent to inequality (149) with a single heat bath, because their setting are mathematically different. However, their physical meanings are physically the same; the both describe the Clausius inequality in the presence of a single heat bath, where entropy S is identified to the von Neumann entropy. We note that

$$\sum_{n=1}^{N-1}(\text{tr}[\rho_n \ln \rho_{ss,n+1}] - \text{tr}[\rho_n \ln \rho_{ss,n}])$$
$$= \beta(\Delta F - \langle W \rangle), \qquad (207)$$

where $\Delta F := F_N - F_1$ is the free-energy difference, and

$$\langle W \rangle := \sum_{n=1}^{N-1}(\text{tr}[\rho_n H_{n+1}] - \text{tr}[\rho_n H_n]) \qquad (208)$$

is the work performed on the system. If the initial and final states are the canonical distributions such that $\rho_0 = \rho_{ss,1}$ and $\rho_N = \rho_{ss,N}$, inequality (204) leads to

$$\langle W \rangle \geq \Delta F. \qquad (209)$$

We note that inequality (209) still holds when the final state ρ_N is out of equilibrium.

6. Second Law with Quantum Feedback Control

We next discuss a generalization of the second law of thermodynamics with quantum feedback control,[72–86] based on the positivity and the monotonicity parallel to Sec. 5. Here, quantum feedback control[101] means that the control protocol on a system depends on an outcome of a quantum measurement on the system. The obtained generalization includes the QC-mutual information. We note that any derivation based on the quantum fluctuation theorem has not been know so far except for a special case,[83,85] while the generalized second law with classical feedback control has been obtained from a generalized fluctuation theorem.[102–107]

6.1. From the Positivity of the Quantum Relative Entropy

We first derive a generalization of Theorem 5.2 with quantum feedback control. Let $\rho_i \in Q(H)$ be the initial state of the system. We first perform a unitary operation U so that the system evolves to $\rho' := U\rho_i U^\dagger$. We next perform a quantum measurement with the Kraus operators $\{M_b\}_{b \in B}$ and the POVM $\{E_b\}_{b \in B}$ with

$$E_b = M_b^\dagger M_b, \tag{210}$$

which is an important assumption in this section. The probability of outcome b is $p(b) = \text{tr}[E_b \rho']$, and the post-measurement state with outcome b is $\rho'(b) := M_b \rho' M_b^\dagger / p(b)$. The QC-mutual information for this measurement is given by

$$I_{\text{QC}} = S(\rho') - \sum_b p(b) S(\rho'(b)). \tag{211}$$

We then perform a unitary operation U_b that depends on outcome b, which is the feedback control. The final state with outcome b is $\rho_f(b) := U_b \rho'(b) U_b^\dagger$, whose ensemble average is $\rho_f := \sum_b U_b M_b U \rho_i U^\dagger M_b^\dagger U_b^\dagger$. In this setup, Theorem 5.2 is generalized to the following theorem.

Theorem 6.1. *Let $\rho_0(b)$ be a reference state corresponding to outcome b. Then*

$$-\sum_b p(b)\text{tr}[\rho_f(b) \ln \rho_0(b)] - S(\rho_i) = \sum_b p(b) S(\rho_f(b) \| \rho_0(b)) - I_{\text{QC}} \tag{212}$$

holds, which leads to

$$-\sum_b p(b)\text{tr}[\rho_f(b) \ln \rho_0(b)] - S(\rho_i) \geq -I_{\text{QC}}. \tag{213}$$

Proof. We straightforwardly calculate that

$$\sum_b p(b) S(\rho_f(b) \| \rho_0(b))$$

$$= -\sum_b p(b) \left(S(\rho_f(b)) + \mathrm{tr}[\rho_f(b) \ln \rho_0(b)] \right)$$

$$= -\sum_b p(b) \mathrm{tr}[\rho_f(b) \ln \rho_0(b)] - S(\rho_i) + S(\rho_i) - \sum_b p(b) S(\rho_f(b)) \quad (214)$$

$$= -\sum_b p(b) \mathrm{tr}[\rho_f(b) \ln \rho_0(b)] - S(\rho_i) + S(\rho') - \sum_b p(b) S(\rho'(b))$$

$$= -\sum_b p(b) \mathrm{tr}[\rho_f(b) \ln \rho_0(b)] - S(\rho_i) + I_{QC},$$

which implies Eq. (212). $\qquad\qquad\square$

Example 6.1. We apply the setup of Example 5.1 to the above theorem. Let H_i and $H_f(b)$ be the initial and final Hamiltonians, where the final one can depend on outcome b. We assume that the initial state is the canonical distribution $\rho_i = e^{\beta(F_i - H_i)}$ with $F_i := -\beta^{-1} \ln \mathrm{tr}[e^{-\beta H_i}]$, and that the reference state is also the canonical distribution $\rho_0(b) = e^{\beta(F_f(b) - H_f(b))}$ with $F_f(b) := -\beta^{-1} \ln \mathrm{tr}[e^{-\beta H_f(b)}]$. We then have

$$-\sum_b p(b) \mathrm{tr}[\rho_f(b) \ln \rho_0(b)] - S(\rho_i) = \beta \langle W - \Delta F \rangle, \quad (215)$$

where

$$\langle W \rangle := \sum_b p(b) \mathrm{tr}[H_f(b) \rho_f(b)] - \mathrm{tr}[H_i \rho_i] \quad (216)$$

is the average of the work performed on the system, and

$$\langle \Delta F \rangle := \sum_b p(b) F_f(b) - F_i \quad (217)$$

is the average of the free-energy difference. Here, we assumed that the energy change of the system during the measurement is the work. Physically, this assumption implies that the measurement process is adiabatic. Inequality (213) then reduces to

$$\beta \langle W - \Delta F \rangle \geq -I_{QC}, \quad (218)$$

which is the generalization of inequality (139) to feedback-controlled processes.

Inequality (218) identifies the upper bound of the sum of the extractable work $-\langle W \rangle$ and the free-energy gain $\langle \Delta F \rangle$, which is proportional to the

QC-mutual information as $\beta^{-1}I_{\text{QC}}$. In the classical limit, the upper bound is given by the classical mutual information. The equality in (218) is achieved by a classical model called the Szilard engine,[108] where $I_{\text{QC}} = \ln 2$, $\langle W \rangle = \beta^{-1}\ln 2$, and $\langle \Delta F \rangle = 0$. Several models that achieves the equality in (218) have been discussed for both classical[107,109,110] and quantum[80] regimes. We note that inequality (218) was first obtained by Sagawa and Ueda.[79]

We note that, in a similar manner to the above example, we can apply the setup of Example. 5.2 to Theorem 6.1, and obtain

$$\Delta S - \sum_k \beta_k \langle Q_k \rangle \geq -I_{\text{QC}}. \tag{219}$$

6.2. From the Monotonicity of the Quantum Relative Entropy

We next derive a generalization of the quantum Hatano-Sasa inequality (201) (Theorem 5.5) with quantum feedback control, based on the monotonicity of the quantum relative entropy. Let $\rho_i \in Q(\boldsymbol{H})$ be the initial state. We control the system through external parameters without feedback. The system then evolves by a CPTP map $\mathcal{E}_m \circ \cdots \circ \mathcal{E}_2 \circ \mathcal{E}_1$, where each $\mathcal{E}_n$ $(n = 1, 2, \cdots, m)$ has a unique steady state $\rho_{\text{ss},n}$. We define

$$\sigma_{\text{ex}}^B(1 \to m) := \sum_{n=1}^{m} \left(\text{tr}[\rho_n \ln \rho_{\text{ss},n}] - \text{tr}[\rho_{n-1} \ln \rho_{\text{ss},n}] \right), \tag{220}$$

where $\rho_n := \mathcal{E}_n(\rho_{n-1})$ and $\rho_0 := \rho_i$.

We next perform a measurement on ρ_m with instrument $\{\mathcal{E}_{\text{meas}}^{(b)}\}$, which is assumed to be given by

$$\mathcal{E}_{\text{meas}}^{(b)}(\rho_m) := M_b \rho_m M_b^\dagger, \tag{221}$$

where M_b's are Kraus operators and b's are measurement outcomes. We note that $\sum_b \mathcal{E}_{\text{meas}}^{(b)}$ is trace-preserving and that $\sum_b M_b^\dagger M_b = I$ holds with the identity $I \in L(\boldsymbol{H})$. The corresponding POVM $\{E_b\}$ satisfies $E_b = M_b^\dagger M_b$ for any b. The probability of obtaining outcome b is $p(b) = \text{tr}[E_b \rho_m]$, and the post-measurement state with outcome b is $\rho_m(b) := M_b \rho_m M_b^\dagger / p(b)$. The QC-mutual information for this measurement is given by

$$I_{\text{QC}} = S(\rho_m) - \sum_b p(b) S(\rho_m(b)). \tag{222}$$

We define $\sigma_{\text{ex}}^B = 0$ during the measurement. Physically, this definition implies that the measurement process is assumed to be adiabatic, which is consistent with the assumption of the adiabaticity in Example 6.1.

We next perform feedback control with a CPTP map $\mathcal{E}_N^{(b)} \circ \cdots \circ \mathcal{E}_{m+2}^{(b)} \circ \mathcal{E}_{m+1}^{(b)}$ that depends on outcome b. We assume that each CPTP map $\mathcal{E}_n^{(b)}$ $(n = m+1, \cdots, N)$ has a unique steady state $\rho_{\mathrm{ss},n}^{(b)}$. We define

$$\sigma_{\mathrm{ex}}^B(m+1 \to N; b) := \sum_{n=m+1}^N \left(\mathrm{tr}[\rho_n^{(b)} \ln \rho_{\mathrm{ss},n}^{(b)}] - \mathrm{tr}[\rho_{n-1}^{(b)} \ln \rho_{\mathrm{ss},n}^{(b)}] \right), \quad (223)$$

where $\rho_n^{(b)} := \mathcal{E}_n^{(b)}(\rho_{n-1}^{(b)})$ $(n = m+1, \cdots, N)$, and define

$$\sigma_{\mathrm{ex}}^B := \sigma_{\mathrm{ex}}^B(1 \to m) + \sum_b p(b)\sigma_{\mathrm{ex}}^B(m+1 \to N; b). \quad (224)$$

We note that the ensemble average of the time evolution is given by

$$\mathcal{E} = \sum_b \mathcal{E}_N^{(b)} \circ \cdots \circ \mathcal{E}_{m+1}^{(b)} \circ \mathcal{E}_{\mathrm{meas}}^{(b)} \circ \mathcal{E}_m \circ \cdots \circ \mathcal{E}_2 \circ \mathcal{E}_1, \quad (225)$$

which is CPTP. We write $\rho_{\mathrm{f}} := \mathcal{E}(\rho_{\mathrm{i}}) = \sum_b p(b)\rho_N^{(b)}$. Then, the quantum Hatano-Sasa inequality (201) is generalized to the following theorem.

Theorem 6.2. *In the above setup,*

$$\sum_b p(b)S(\rho_N^{(b)}) - S(\rho_{\mathrm{i}}) \geq -\sigma_{\mathrm{ex}}^B - I_{\mathrm{QC}}. \quad (226)$$

Proof. We have a set of inequalities from the monotonicity of the relative entropy:

$$S(\rho_{n+1}) - S(\rho_n) \geq -\left(\mathrm{tr}[\rho_n \ln \rho_{\mathrm{ss},n}] - \mathrm{tr}[\rho_{n-1} \ln \rho_{\mathrm{ss},n}]\right) \ (1 \leq n \leq m), \quad (227)$$

$$S(\rho_{n+1}^{(b)}) - S(\rho_n^{(b)}) \geq \left(\mathrm{tr}[\rho_n^{(b)} \ln \rho_{\mathrm{ss},n}^{(b)}] - \mathrm{tr}[\rho_{n-1}^{(b)} \ln \rho_{\mathrm{ss},n}^{(b)}]\right) \ (m+1 \leq n \leq N). \quad (228)$$

By combining them to Eq. (222), we obtain inequality (226). $\square$

Inequality (226) has been obtained in this article for the first time. On the other hand, the classical Hatano-Sasa equality and inequality have been generalized to feedback-controlled classical systems.[84,85]

Example 6.2. We apply the setup in Example 5.7 to Theorem 6.2. Let H_n be the Hamiltonian corresponding to $\mathcal{E}_n$ and $F_n := -\beta^{-1} \ln \mathrm{tr}[e^{-\beta H_n}]$ be the free energy $(1 \leq n \leq N)$. We note that H_n and F_n can depend on outcome b for $n \geq m+1$, which are denoted as $H_n(b)$ and $F_n(b)$. We assume that the steady states are the canonical distributions: $\rho_{ss,n} = e^{\beta(F_n - H_n)}$ $(n \leq m)$ and $\rho_{ss,n}^{(b)} = e^{\beta(F_n(b) - H_n(b))}$ $(n \geq m + 1)$. In this case, σ_{ex}^B is given by

$$
\begin{aligned}
\sigma_{ex}^B = & -\beta \sum_{n=1}^{m} (\text{tr}[H_n \rho_n] - \text{tr}[H_n \rho_{n-1}]) \\
& -\beta \sum_b p(b) \sum_{n=m+1}^{N} (\text{tr}[H_n(b)\rho_n(b)] - \text{tr}[H_n(b)\rho_{n-1}(b)]) \\
=: & -\beta\langle Q \rangle,
\end{aligned}
\tag{229}
$$

where $\langle Q \rangle$ is the average of the heat absorbed by the system. Here, we assumed that the measurement process is adiabatic such that $\langle Q \rangle = 0$ holds during the measurement, which is consistent with the definition $\sigma_{ex}^B = 0$ during $\mathcal{E}_{meas}^{(b)}$. Inequality (226) then reduces to

$$
\langle \Delta S \rangle \geq \beta \langle Q \rangle - I_{QC},
\tag{230}
$$

where $\langle \Delta S \rangle := \sum_b p(b) S(\rho_N^{(b)}) - S(\rho_i)$.

We further assume that the initial and final states are the canonical distributions such that $\rho_i = \rho_{ss,1}$ and $\rho_N^{(b)} = \rho_{ss,N}^{(b)}$. We then have

$$
\langle \Delta S \rangle = \beta \langle \Delta E - \Delta F \rangle,
\tag{231}
$$

where

$$
\langle \Delta E \rangle := \sum_b p(b) \text{tr}[H_N^{(b)} \rho_N^{(b)}] - \text{tr}[H_1 \rho_i]
\tag{232}
$$

is the energy difference of the system, and $\langle \Delta F \rangle := \sum_b p(b) F_N(b) - F_1$. Therefore, we obtain from inequality (230) that

$$
\beta \langle W - \Delta F \rangle \geq -I_{QC},
\tag{233}
$$

where we defined $\langle W \rangle := \langle \Delta E \rangle - \langle Q \rangle$ due to the first law of thermodynamics. While inequality (233) is mathematically not equivalent to inequality (218), their physical meanings are the same. We note that inequality (233) still holds when $\rho_N^{(b)}$'s are out of equilibrium.

7. Concluding Remarks

We have discussed several second law-like inequalities based on the positivity and the monotonicity of the quantum relative entropy. In Sec. 2, we discussed basic concepts for quantum states and dynamics. In particular, we discussed the complete positivity of time evolutions, and derived the Kraus representation (20). In Sec. 3, we introduced the von Neumann entropy and the quantum relative entropy, and discussed their basic properties. In particular, we proved the positivity of the quantum relative entropy. We also discussed the monotonicity, which is proved in Appendix B. In Sec. 4, we have discussed the quantum mutual information and two related quantities. One is the Holevo's χ-quantity and the other is the QC-mutual information. The two quantities characterize the accessible classical information encoded in quantum states, through the Holevo bound (102) and its dual inequality (118). We also showed the data processing inequalities (77) and (98), which are direct consequences of the monotonicity of the quantum relative entropy.

Sections 5 is devoted to the derivations of the second law of thermodynamics and its generalization. In Sec. 5.1, we discussed that the von Neumann entropy and the thermodynamic entropy can be identified in canonical distributions. In Sec. 5.2, we derived a general inequality (130) based on the positivity. Inequality (130) leads to a well-known expression of the second law of thermodynamics (149) (i.e., the Clausius inequality) as a special case, by treating the total system including the baths as a unitary system. In Sec. 5.3, we discussed the quantum fluctuation theorem, which leads to the second law of thermodynamics (168) that is equivalent to inequality (130) in Sec. 5.2. In Sec. 5.4, we derived a general inequality (201) based on the monotonicity, which is regarded as a quantum version of the Hatano-Sasa inequality. This inequality leads to the second law of thermodynamics (206) for a special case. Our derivations are independent of the size of the system, and therefore the obtained inequalities can be applied to small thermodynamic systems such as quantum dots.

We now discuss the physical meanings of the results in more detail. We first compare the two derivations of the second law, which are based on the positivity and the monotonicity. The derivation of inequality (149) based on the positivity is quite universal, because we have essentially made only three assumptions: (i) the total system including the baths obeys a unitary evolution, (ii) the baths are initially in the canonical distributions, and (iii) the system is initially not correlated to the baths. We stress that we did not assume anything about the intermediate and the final states of the total system. The assumptions (i) and (ii) would be physically reasonable. The assumption (iii) can be justified if the system is initially separated from the baths.

On the other hand, in the derivation of inequality (198) based on the monotonicity, any bath is not explicitly included in our formulation, and therefore time evolutions are assumed to be nonunitary. In this case, the derivation requires a strong assumption that (iv) the steady state of each CPTP map $\mathcal{E}_n$ is a canonical distribution. We stress that the assumption (iv) is not satisfied in general if we trace out the bath from the total system satisfying (i)-(iii). Moreover, the derivation based on the monotonicity implicitly requires that (v) the outside bath needs to be refreshed (i.e., needs to be replaced by a new bath that is not correlated with the system) when the external parameters are changed, because each $\mathcal{E}_n$ is assumed to be CPTP.

We note that, in the presence of more than one baths, the monotonicity cannot be applied to derive the Clausius inequality. On the other hand, the monotonicity can be applied to the cases in which the steady states are not in thermal equilibrium, which is an advantage of the method of the monotonicity. We note that the derivation based on the monotonicity can be generalized to continuous-time systems described by quantum master equations.[59,101,111,112]

Going back to the assumption (iii) for the derivation based on the positivity, it has been pointed out that if the system is initially correlated with the baths, inequality (149) needs modifications.[113–115] For example, in the special case of inequality (156), the free energy of the system needs to be renormalized.[47] On the other hand, the role of the initial correlation has been discussed in terms of the origin of the arrow of the time.[116,117]

The assumption (ii) can be criticized in terms of the foundation of statistical mechanics. In macroscopic systems, a thermal equilibrium state does not necessarily correspond to any canonical distribution from the microscopic point of view. In fact, it has been shown that even a pure state can behave as a thermal equilibrium state.[118–126] Therefore, the assumption (ii) is difficult to be rigorously justified for macroscopic baths. The identification of the von Neumann entropy to the thermodynamic entropy fails in general, except for the cases in which a thermal equilibrium state corresponds to a canonical distribution. Experimentally, trapped ultracold atoms can relax to a thermal equilibrium state even if they are well separated from the environment. We note that the detailed investigations of the relaxation processes of isolated quantum systems are now experimentally accessible.[127]

We next briefly mention a theoretical approach to derive the second law without assuming the initial canonical distribution.[128–131] Let $\rho \in Q(\boldsymbol{H})$ be an initial state. We consider a unitary evolution U from $t = 0$ to τ with a time-dependent Hamiltonian $H(t)$. We define the work performed on the system as $\langle W \rangle := \mathrm{tr}[H(\tau)U\rho U^{\dagger}] - \mathrm{tr}[H(0)\rho]$. We also define the quasi-static work $\langle W \rangle_{\text{slow}} := \lim_{\tau \to \infty} \langle W \rangle$, where $H(0)$ and $H(\tau)$ are fixed. Let $H(0) = \sum_k E_k |\psi_k\rangle\langle\psi_k|$ be the spectrum decomposition of the initial Hamiltonian. The crucial assumption on the initial state is that ρ is written as $\rho = \sum_k p(k)|\psi_k\rangle\langle\psi_k|$ with $p(k)$'s satisfying

$$p(k) \geq p(k') \quad \text{if} \quad E_k \leq E_{k'}. \tag{234}$$

Then, Lenard[128] proved that

$$\langle W \rangle \geq \langle W \rangle_{\text{slow}}, \tag{235}$$

which is regarded as an expression of the second law of thermodynamics. We note that the quasi-static work can be identified to the free-energy difference as $\langle W \rangle_{\text{slow}} = \Delta F$.

Without assuming (234), inequality (235) is not satisfied in general. For example, the microcanonical distribution does not satisfy assumption (234), if the entire Hilbert space is spanned by all $|\psi_k\rangle$'s and is not restricted to the microcanonical energy shell. In fact, several counterexamples against inequality (235) have been discussed in classical systems with the microcanonical initial distribution.[132–134] We note that a generalized quantum fluctuation theorem for microcanonical initial distributions has been discussed.[135]

The role of a quantum coherence in the initial state has been studied in terms of quantum heat engines.[136–142] For example, if the initial state differs from the canonical distribution and involves an additional quantum coherence, the second law of thermodynamics like (235) is not necessarily satisfied.

We now reach an observation that the choice of the initial state is crucial to derive the second law. In fact, the canonical distribution is a special state that maximizes its von Neumann entropy among the states that have the same amount of the energy. This special initial condition is a crucial reason why we can derive the second law even if the dynamics of the total system is reversible. It would be difficult to rigorously justify the physical validity of the initial canonical distribution. Therefore, the derivations of the second law in Sec. 5 are not so satisfactory to understand the fundamental reason why the macroscopic world is irreversible in spite of the reversibility of the microscopic dynamics. To understand the origin of the arrow of the time would be an interesting but quite difficult future challenge.

On the other hand, thermodynamics of information processing is also an interesting topic. While it would be easier than the aforementioned fundamental problem, it is also closely related to the foundation of thermodynamics and statistical mechanics. This topic has been discussed by numerous researchers in terms of the paradox of "Maxwell's demon,"[144-147] which can be formulated as an information processing device. It has been understood that the demon is consistent with the conventional second law of thermodynamics.[146-149] Recently, modern nonequilibrium statistical mechanics and quantum information theory have shed new light on the theory of thermodynamics including the demon.[72-86,102-107,109,110,144-149]

In particular, we discussed thermodynamics of quantum feedback control in Sec. 6. We obtained inequalities (213) and (226) as generalizations of (130) and (201), respectively. The obtained inequalities lead to generalizations of the second law (218) and (230). We note that the quantum fluctuation theorem with quantum information processing has been elusive except for special cases.[83,85]

Appendix A. Short Summary of the Linear Algebra

In this appendix, we briefly summarize the concepts and notations of the linear algebra of finite-dimensional Hilbert spaces. Let H be a N-dimensional Hilbert space with $N < \infty$. The inner product of $|\varphi\rangle, |\phi\rangle \in H$ is written as $\langle \varphi | \phi \rangle$ or (φ, ϕ). The norm of the Hilbert space is given by $\|\varphi\|^2 = \langle \varphi | \varphi \rangle$. A linear basis $\{|\varphi_k\rangle\}_{k=1}^{N} \subset H$ is called an orthonormal basis if it satisfies $\langle \varphi_k | \varphi_l \rangle = \delta_{kl}$ with δ_{kl} the Kronecker delta.

Let $L(H, H')$ be the set of linear operators from H to another Hilbert space H'. In particular, we write $L(H) := L(H, H)$. We write $\langle \varphi | X | \phi \rangle := (\varphi, X\phi)$ for $X \in L(H)$. For any $X \in L(H, H')$, there exists a unique operator $X^\dagger \in L(H', H)$ that satisfies

$$(\phi, X\psi) = (X^\dagger \phi, \psi) \tag{A.1}$$

for any $|\psi\rangle \in H$ and $|\phi\rangle \in H'$. $X^\dagger$ is called the adjoint or the Hermitian conjugate operator of X. If $X^\dagger = X$ holds for $X \in L(H)$, X is called self-adjoint or Hermitian. We note that any operator $X \in L(H)$ can be written as $X = X_1 + iX_2$, where $X_1 := (X + X^\dagger)/2$ and $X_2 := (X - X^\dagger)/(2i)$ are Hermitian.

Let H' be a subspace of H with dimension N' ($\leq N$). Let $\{|\varphi_k\rangle\}_{k=1}^{N'}$ be an orthonormal basis of H'. The projection operator onto H' is given by

$$P_{H'} = \sum_{k=1}^{N'} |\varphi_k\rangle\langle\varphi_k| \in L(H). \tag{A.2}$$

In particular, any orthonormal basis satisfies

$$\sum_{k=1}^{N} |\varphi_k\rangle\langle\varphi_k| = I, \tag{A.3}$$

where I is the identity operator on $\boldsymbol{H}$. On the other hand, $|\varphi\rangle\langle\varphi|$ with $\|\varphi\| = 1$ is the projection operator onto the 1-dimensional space that is spanned by $|\varphi\rangle$. Any Hermitian operator X has the spectrum decomposition

$$X = \sum_k x_k |\varphi_k\rangle\langle\varphi_k|, \tag{A.4}$$

where $x_k \in \mathbb{R}$ is an eigenvalue and $|\varphi_k\rangle$ is an eigenvector that constitutes an orthonormal basis $\{|\varphi_k\rangle\} \subset \boldsymbol{H}$. The support of X is the subspace of $\boldsymbol{H}$ that is spanned by $|\varphi_k\rangle$'s with nonzero eigenvalues.

If $\langle\varphi|X|\varphi\rangle \geq 0$ holds for any $|\varphi\rangle \in \boldsymbol{H}$, $X \in L(\boldsymbol{H})$ is called positive. Any positive operator is Hermitian, because $\langle\varphi|X_2|\varphi\rangle$ should be zero for any $|\varphi\rangle$ with X_2 defined above. Any eigenvalue of a positive operator is non-negative. We write $X \geq Y$ if $X - Y$ is positive. In particular, $X \geq 0$ if X is positive. If $\langle\varphi|X|\varphi\rangle > 0$ holds for any nonzero $|\varphi\rangle \in \boldsymbol{H}$, $X \in L(\boldsymbol{H})$ is called positive definite. Any positive-definite operator is Hermitian with positive eigenvalues. We note that the support of a positive-definite operator equals to $\boldsymbol{H}$. We also note that $V \in L(\boldsymbol{H}, \boldsymbol{H}')$ is called a contraction, if $\langle\varphi|V^\dagger V|\varphi\rangle \leq \langle\varphi|\varphi\rangle$ holds for any $|\varphi\rangle \in \boldsymbol{H}$.

The trace of $X \in L(\boldsymbol{H})$ is defined as

$$\text{tr}[X] := \sum_k \langle\varphi_k|X|\varphi_k\rangle, \tag{A.5}$$

where $\{|\varphi_k\rangle\} \subset \boldsymbol{H}$ is an orthonormal basis. We note that $\text{tr}[X]$ is independent of the choice of the orthonormal basis. If the spectrum decomposition of X is given by Eq. (A.4), its trace is given by $\text{tr}[X] = \sum_k x_k$. Let $X \in L(\boldsymbol{H}, \boldsymbol{H}')$ and $Y \in L(\boldsymbol{H}', \boldsymbol{H})$. We then have

$$\text{tr}[YX] = \text{tr}[XY], \tag{A.6}$$

where the left and right traces are on $\boldsymbol{H}$ and $\boldsymbol{H}'$, respectively. In fact,

$$\text{tr}[YX] = \sum_{kl} \langle\phi_k|Y|\psi_l\rangle\langle\psi_l|X|\phi_k\rangle = \sum_{kl} \langle\psi_l|X|\phi_k\rangle\langle\phi_k|Y|\psi_l\rangle = \text{tr}[XY], \tag{A.7}$$

where $\{|\phi_k\rangle\}$ and $\{|\psi_k\rangle\}$ are respectively orthonormal bases of $\boldsymbol{H}$ and $\boldsymbol{H}'$.

Let $\boldsymbol{H}$ and $\boldsymbol{H}'$ be Hilbert spaces with dimensions N and N', respectively. For any $X \in L(\boldsymbol{H}, \boldsymbol{H}')$, there exist orthonormal bases $\{|\varphi_k\rangle\}_{k=1}^{N} \subset$

H and $\{|\psi_k\rangle\}_{k=1}^{N'} \subset H'$ such that

$$X = \sum_{k=1}^{N''} \lambda_k |\psi_k\rangle\langle\varphi_k|, \qquad (A.8)$$

where $\lambda_k \geq 0$ $(1 \leq k \leq N'')$ and $N'' := \min\{N, N'\}$. Equality (A.8) is called the singular value decomposition, and λ_k is called a singular value of X.

We next consider the tensor product of two Hilbert spaces. Let H_A and H_B be two Hilbert spaces, and $H_A \otimes H_B$ be their tensor product. For simplicity, we write $|\varphi_A\rangle \otimes |\varphi_B\rangle \in H_A \otimes H_B$ as $|\varphi_A\rangle|\varphi_B\rangle$. Let $\{|\varphi_k\rangle\}$ and $\{|\psi_l\rangle\}$ be orthonormal bases of H_A and H_B, respectively. Any vector $|\Psi\rangle \in H_A \otimes H_B$ can be written as

$$|\Psi\rangle = \sum_{kl} \alpha_{kl} |\varphi_k\rangle|\psi_l\rangle, \qquad (A.9)$$

where $\alpha_{kl} \in \mathbb{C}$. By applying the singular-value decomposition to matrix (α_{kl}), we find that there are orthonormal bases $\{|\varphi_k'\rangle\} \subset H_A$ and $\{|\psi_l'\rangle\} \subset H_B$ such that

$$|\Psi\rangle = \sum_{k} \lambda_k |\varphi_k'\rangle|\psi_k'\rangle, \qquad (A.10)$$

where $\lambda_k \geq 0$ is a singular value of matrix (α_{kl}). Equality (A.10) is called the Schmidt decomposition of $|\Psi\rangle$.

Appendix B. Proof of the Monotonicity of the Quantum Relative Entropy

In this appendix, we prove the monotonicity (64) of the quantum relative entropy (Theorem 3.5) in line with Petz's proof.[100] We first prove some lemmas, in which the key concepts are the operator monotonicity and the operator convexity.[150]

Let H be a finite-dimensional Hilbert space and $X \in L(H)$ be a positive-definite operator with spectrum decomposition $X = \sum_k x_k |\varphi_k\rangle\langle\varphi_k|$. We can substitute X to a function $f : (0, \infty) \to \mathbb{R}$ as

$$f(X) := \sum_{k} f(x_k) |\varphi_k\rangle\langle\varphi_k|. \qquad (B.1)$$

We then introduce the following concepts.

Definition B.1. f is called decreasing-operator monotone, if $f(X) \geq f(Y)$ holds for any positive-definite operators $X, Y \in L(H)$ satisfying $X \leq Y$.

Definition B.2. f is called operator convex, if $f(pX+(1-p)Y) \leq pf(X)+ (1-p)f(Y)$ holds for any $0 \leq p \leq 1$ and any positive-definite operators $X, Y \in L(\boldsymbol{H})$.

In general, it is difficult to judge whether a function is operator monotone and is operator convex. However, it is not so difficult to show that $f(x) = (x+t)^{-1}$ $(t \geq 0)$ and $f(x) = -\ln x$ are both decreasing-operator monotone and operator convex.

Lemma B.1. $f(x) = (x+t)^{-1}$ $(t \geq 0)$ *is decreasing-operator monotone.*

Proof. It is obvious that $X \leq I \Rightarrow X^{-1} \geq I$ with $I \in \boldsymbol{H}$ the identity. We then have $Y^{-1/2}XY^{-1/2} \leq I \Rightarrow Y^{1/2}X^{-1}Y^{1/2} \geq I$, and therefore $X \leq Y \Rightarrow X^{-1} \geq Y^{-1}$. Since $X \leq Y \Rightarrow X+tI \leq Y+tI$, we obtain $X \leq Y \Rightarrow (X+tI)^{-1} \geq (Y+tI)^{-1}$. □

Lemma B.2. $f(x) = (x+t)^{-1}$ $(t \geq 0)$ *is operator convex.*

Proof. Since x^{-1} is convex, $(pX + (1-p)I) \leq pX^{-1} + (1-p)I$ holds. We then have $(pY^{-1/2}XY^{-1/2} + (1-p)I) \leq pY^{1/2}X^{-1}Y^{1/2} + (1-p)I$, and therefore $(pX + (1-p)Y) \leq pX^{-1} + (1-p)Y^{-1}$. By noting that $(pX+(1-p)Y+tI) = (p(X+tI)+(1-p)(Y+tI))$, we obtain $(pX + (1-p)Y + tI)^{-1} \leq p(X+tI)^{-1} + (1-p)(Y+tI)^{-1}$. □

Lemma B.3. $f(x) := -\ln x$ *is decreasing-operator monotone and operator convex.*

Proof. It follows from that $-\ln x = \int_0^\infty \left((x+t)^{-1} - (1+t)^{-1}\right) dt$. □

It is known that any decreasing-operator monotone function is operator convex.[151] We next show an important property of operator convex functions.

Lemma B.4 (Jensen's operator inequality[152-154]). *If f is operator convex,*

$$f(V^\dagger X V) \leq V^\dagger f(X) V \tag{B.2}$$

holds for any contraction $V \in L(\boldsymbol{H}, \boldsymbol{H}')$ and any positive-definite operator $X \in L(\boldsymbol{H}')$ such that $V^\dagger X V$ is also positive definite.[e]

---

[e]The definition of a contraction is given in Appendix A.

Proof. We define

$$X' := \begin{bmatrix} X & 0 \\ 0 & 0 \end{bmatrix} \in L(\boldsymbol{H}' \oplus \boldsymbol{H}, \boldsymbol{H}' \oplus \boldsymbol{H}),$$

$$V_1 := \begin{bmatrix} V & (I - VV^\dagger)^{1/2} \\ (I - V^\dagger V)^{1/2} & -V^\dagger \end{bmatrix} \in L(\boldsymbol{H} \oplus \boldsymbol{H}', \boldsymbol{H}' \oplus \boldsymbol{H}), \qquad (B.3)$$

$$V_2 := \begin{bmatrix} V & -(I - VV^\dagger)^{1/2} \\ (I - V^\dagger V)^{1/2} & V^\dagger \end{bmatrix} \in L(\boldsymbol{H} \oplus \boldsymbol{H}', \boldsymbol{H}' \oplus \boldsymbol{H}),$$

where we used the assumption that V is a contraction to define $(I - V^\dagger V)^{1/2}$ and $(I - VV^\dagger)^{1/2}$. By noting that $\boldsymbol{H} \oplus \boldsymbol{H}' \simeq \boldsymbol{H}' \oplus \boldsymbol{H}$ and by using the singular-value decomposition of V, it is easy to check that V_1 and V_2 are unitary. We have

$$V_1^\dagger X' V_1 = \begin{bmatrix} V^\dagger X V & V^\dagger X U \\ U X V & U X U \end{bmatrix}, \quad V_2^\dagger X' V_2 = \begin{bmatrix} V^\dagger X V & -V^\dagger X U \\ -U X V & U X U \end{bmatrix}, \qquad (B.4)$$

where $U := (I - VV^\dagger)^{1/2}$, and

$$\frac{V_1^\dagger X' V_1 + V_2^\dagger X' V_2}{2} = \begin{bmatrix} V^\dagger X V & 0 \\ 0 & U X U \end{bmatrix}. \qquad (B.5)$$

Therefore, by using the operator convexity of f, we obtain

$$\begin{bmatrix} f(V^\dagger X V) & 0 \\ 0 & f(U X U) \end{bmatrix} = f\left(\frac{V_1^\dagger X' V_1 + V_2^\dagger X' V_2}{2} \right) \leq \frac{f(V_1^\dagger X' V_1) + f(V_2^\dagger X' V_2)}{2}$$

$$= \frac{V_1^\dagger f(X') V_1 + V_2^\dagger f(X') V_2}{2} = \begin{bmatrix} V^\dagger f(X) V & 0 \\ 0 & U f(X) U \end{bmatrix},$$

$$\qquad (B.6)$$

which implies inequality (B.2). $\square$

Conversely, it is known that[153] if inequality (B.2) is satisfied for any contraction $V \in L(\boldsymbol{H}, \boldsymbol{H}')$ and any positive-definite operator $X \in L(\boldsymbol{H}')$, then f is operator convex.

We next consider a generalization of the Schwarz inequality.

Lemma B.5 (Kadison inequality[155]). *Let $X \in \boldsymbol{H}$ be a Hermitian operator and $\mathcal{E} : L(\boldsymbol{H}) \to L(\boldsymbol{H}')$ be a unital positive map.[f] Then*

$$\mathcal{E}(X^2) \geq \mathcal{E}(X)^2. \qquad (B.7)$$

---

[f] The definition of a unital map is given in Sec. 2.2.2.

Proof. Let $X := \sum_{k=1}^{N} x_k |\varphi_k\rangle\langle\varphi_k|$ be the spectrum decomposition of X, where N is the dimension of $\boldsymbol{H}$. We define $X_k := \mathcal{E}(|\varphi_k\rangle\langle\varphi_k|)$, which satisfies $X_k \geq 0$ due to the positivity of $\mathcal{E}$ and $\sum_k X_k = I$ due to assumption $\mathcal{E}(I) = I$. Inequality (B.7) can then be written as

$$\sum_k x_k^2 X_k \geq \left(\sum_k x_k X_k\right)^2, \qquad (B.8)$$

or equivalently, for any $|\psi\rangle \in \boldsymbol{H}'$,

$$\sum_k \langle\psi| x_k^2 X_k |\psi\rangle \geq \langle z|z\rangle, \qquad (B.9)$$

where $|z\rangle := \sum_k x_k X_k |\psi\rangle$. We introduce auxiliary system $\mathbb{C}^N$ that has an orthonormal basis $\{|e_k\rangle\}_{k=1}^N$, and define an inner product $(\cdot,\cdot)_K$ in $\boldsymbol{H}'\otimes\mathbb{C}^N$ such that

$$\left(\sum_k |\varphi_k\rangle|e_k\rangle, \sum_k |\psi_k\rangle|e_k\rangle\right)_K := \sum_k \langle\varphi_k|X_k|\psi_k\rangle. \qquad (B.10)$$

By using the Schmidt inequality for the inner product $(\cdot,\cdot)_K$, we have

$$\langle z|z\rangle = \sum_k \langle z| x_k X_k |\psi\rangle$$

$$= \left(\sum_k |z\rangle|e_k\rangle, \sum_k x_k|\psi\rangle|e_k\rangle\right)_K$$

$$\leq \left(\sum_k |z\rangle|e_k\rangle, \sum_k |z\rangle|e_k\rangle\right)_K^{1/2} \left(\sum_k x_k|\psi\rangle|e_k\rangle, \sum_k x_k|\psi\rangle|e_k\rangle\right)_K^{1/2}$$

$$= \left(\sum_k \langle z|X_k|z\rangle\right)^{1/2} \left(\sum_k \langle\psi| x_k X_k x_k |\psi\rangle\right)^{1/2}$$

$$= \langle z|z\rangle^{1/2} \left(\sum_k \langle\psi| x_k^2 X_k |\psi\rangle\right)^{1/2},$$

$$(B.11)$$

which implies inequality (B.9). $\qquad\square$

We note that we did not assume the complete positivity of $\mathcal{E}$ for the Kadison inequality. The following lemma is a straightforward consequence:

Lemma B.6 (Schwarz's operator inequality[156]). *Let $X \in L(H)$ be an arbitrary operator and $\mathcal{E} : L(H) \to L(H')$ be a unital 2-positive map. Then*

$$\mathcal{E}(X^\dagger X) \geq \mathcal{E}(X^\dagger)\mathcal{E}(X). \tag{B.12}$$

Proof. By applying the Kadison inequality (B.7) to positive map $\mathcal{E} \otimes \mathcal{I}_2$ and a Hermitian operator

$$X' := \begin{bmatrix} 0 & X^\dagger \\ X & 0 \end{bmatrix} \in L(H \otimes \mathbb{C}^2), \tag{B.13}$$

we obtain $(\mathcal{E} \otimes \mathcal{I}_2)(X'^2) \geq ((\mathcal{E} \otimes \mathcal{I}_2)(X'))^2$, or equivalently

$$\begin{bmatrix} \mathcal{E}(X^\dagger X) & 0 \\ 0 & \mathcal{E}(XX^\dagger) \end{bmatrix} \geq \begin{bmatrix} \mathcal{E}(X^\dagger)\mathcal{E}(X) & 0 \\ 0 & \mathcal{E}(X)\mathcal{E}(X^\dagger) \end{bmatrix}, \tag{B.14}$$

which implies inequality (B.12). □

We now prove the monotonicity (64) in Theorem 3.5. Let $\mathcal{E} : L(H) \to L(H')$ be a CPTP map and $\rho, \sigma \in Q(H)$ be states. For simplicity, we assume that ρ, σ, $\mathcal{E}(\rho)$, and $\mathcal{E}(\sigma)$ are positive definite. We define $\mathcal{L}$, $\mathcal{R}$, and $\mathcal{D}$ that act on $L(H)$ such that, for $X \in L(H)$,

$$\mathcal{L}(X) := \sigma X, \ \mathcal{R}(X) := X\rho^{-1}, \ \mathcal{D}(X) := \sigma X\rho^{-1}, \tag{B.15}$$

where $\mathcal{D} = \mathcal{L}\mathcal{R} = \mathcal{R}\mathcal{L}$. We note that $\mathcal{L}$, $\mathcal{R}$, and $\mathcal{D}$ are positive definite in terms of the Hilbert-Schmidt inner product (43).

Let $\sigma := \sum_k p_k |\varphi_k\rangle\langle\varphi_k|$ be the spectrum decomposition. Then the eigenvectors of $\mathcal{L}$ are $\{|\varphi_k\rangle\langle\varphi_l|\}_{kl}$ and the eigenvalues are $\{p_k\}_k$. We then have $(\ln \mathcal{L})(X) = (\ln \sigma)X$. Similarly, $(\ln \mathcal{R})(X) = -X(\ln \rho)$. Therefore,

$$(\ln \mathcal{D})(X) = (\ln \mathcal{L} + \ln \mathcal{R})(X) = (\ln \sigma)X - X(\ln \rho). \tag{B.16}$$

By using the Hilbert-Schmidt inner product, we have

$$\begin{aligned} S(\rho\|\sigma) &= \langle \rho^{1/2}, (\ln \rho)\rho^{1/2}\rangle_{\text{HS}} - \langle \rho^{1/2}, (\ln \sigma)\rho^{1/2}\rangle_{\text{HS}} \\ &= \langle \rho^{1/2}, (-\ln \mathcal{D})\rho^{1/2}\rangle_{\text{HS}}. \end{aligned} \tag{B.17}$$

On the other hand,

$$S(\mathcal{E}(\rho)\|\mathcal{E}(\sigma)) = \langle \mathcal{E}(\rho)^{1/2}, (-\ln \mathcal{D}')(\mathcal{E}(\rho)^{1/2})\rangle_{\text{HS}}, \tag{B.18}$$

where $\mathcal{D}'$ acts on $L(H')$ such that

$$\mathcal{D}'(X) := \mathcal{E}(\sigma)X\mathcal{E}(\rho)^{-1} \ (X \in L(H')). \tag{B.19}$$

We next define $\mathcal{V} : L(\boldsymbol{H}') \to L(\boldsymbol{H})$ such that

$$\mathcal{V}(X) := \mathcal{E}^\dagger \left(X\mathcal{E}(\rho)^{-1/2} \right) \rho^{1/2}, \tag{B.20}$$

or equivalently

$$\mathcal{V}\left(X\mathcal{E}(\rho)^{1/2} \right) = \mathcal{E}^\dagger(X)\rho^{1/2} \tag{B.21}$$

for $X \in L(\boldsymbol{H}')$. We note that $\mathcal{V}(\mathcal{E}(\rho)^{1/2}) = \rho^{1/2}$ holds, because $\mathcal{E}$ is trace-preserving so that $\mathcal{E}^\dagger$ is unital. We then have

$$\begin{aligned}
S(\rho\|\sigma) &= \langle \mathcal{V}(\mathcal{E}(\rho)^{1/2}), (-\ln\mathcal{D})\mathcal{V}(\mathcal{E}(\rho)^{1/2}) \rangle_{\mathrm{HS}} \\
&= \langle \mathcal{E}(\rho)^{1/2}, \mathcal{V}^\dagger(-\ln\mathcal{D})\mathcal{V}(\mathcal{E}(\rho)^{1/2}) \rangle_{\mathrm{HS}}.
\end{aligned} \tag{B.22}$$

Therefore, our goal is to show that

$$\langle \mathcal{E}(\rho)^{1/2}, \mathcal{V}^\dagger(-\ln\mathcal{D})\mathcal{V}(\mathcal{E}(\rho)^{1/2}) \rangle_{\mathrm{HS}} \geq \langle \mathcal{E}(\rho)^{1/2}, (-\ln\mathcal{D}')(\mathcal{E}(\rho)^{1/2}) \rangle_{\mathrm{HS}}. \tag{B.23}$$

To show inequality (B.23), it is enough to show that

$$\mathcal{V}^\dagger(-\ln\mathcal{D})\mathcal{V} \geq -\ln\mathcal{D}'. \tag{B.24}$$

We then show that $\mathcal{V}$ is a contraction in terms of the Hilbert-Schmidt inner product. In fact,

$$\begin{aligned}
\langle \mathcal{E}^\dagger(X)\rho^{1/2}, \mathcal{E}^\dagger(X)\rho^{1/2} \rangle_{\mathrm{HS}} &= \mathrm{tr}\left[\rho\mathcal{E}^\dagger(X^\dagger)\mathcal{E}^\dagger(X) \right] \\
&\leq \mathrm{tr}\left[\rho\mathcal{E}^\dagger(X^\dagger X) \right] = \mathrm{tr}\left[\mathcal{E}(\rho)X^\dagger X \right] \\
&= \langle X\mathcal{E}(\rho)^{1/2}, X\mathcal{E}(\rho)^{1/2} \rangle_{\mathrm{HS}},
\end{aligned} \tag{B.25}$$

where we used the Schwarz inequality (B.12) (Lemma B.6) for $\mathcal{E}^\dagger$ that is unital. Since $-\ln x$ is operator convex from Lemma B.3, we obtain that

$$-\ln(\mathcal{V}^\dagger\mathcal{D}\mathcal{V}) \leq \mathcal{V}^\dagger(-\ln\mathcal{D})\mathcal{V}, \tag{B.26}$$

where we used the Jensen's operator inequality (B.2) (Lemma B.4).

We next show that

$$\mathcal{V}^\dagger\mathcal{D}\mathcal{V} \leq \mathcal{D}'. \tag{B.27}$$

In fact,

$$\begin{aligned}
&\langle X\mathcal{E}(\rho)^{1/2}, \mathcal{V}^\dagger\mathcal{D}\mathcal{V}(X\mathcal{E}(\rho)^{1/2}) \rangle_{\mathrm{HS}} \\
&= \langle \mathcal{V}(X\mathcal{E}(\rho)^{1/2}), \mathcal{D}\mathcal{V}(X\mathcal{E}(\rho)^{1/2}) \rangle_{\mathrm{HS}} \\
&= \langle \mathcal{E}^\dagger(X)\rho^{1/2}, \mathcal{D}(\mathcal{E}^\dagger(X)\rho^{1/2}) \rangle_{\mathrm{HS}} = \mathrm{tr}\left[\rho\mathcal{E}^\dagger(X^\dagger)\mathcal{D}\mathcal{E}^\dagger(X) \right] \\
&= \mathrm{tr}\left[\sigma\mathcal{E}^\dagger(X)\mathcal{E}^\dagger(X^\dagger) \right] \leq \mathrm{tr}\left[\sigma\mathcal{E}^\dagger(XX^\dagger) \right] = \mathrm{tr}\left[\mathcal{E}(\sigma)XX^\dagger \right] \\
&= \mathrm{tr}[\mathcal{E}(\rho)^{1/2}X^\dagger\mathcal{E}(\sigma)X\mathcal{E}(\rho)^{1/2}\mathcal{E}(\rho)^{-1}] \\
&= \langle X\mathcal{E}(\rho)^{1/2}, \mathcal{D}'(X\mathcal{E}(\rho)^{1/2}) \rangle_{\mathrm{HS}},
\end{aligned} \tag{B.28}$$

where we again used the Schwarz inequality (B.12) (Lemma B.6) for $\mathcal{E}^\dagger$. Since $-\ln x$ is decreasing-operator monotone from Lemma B.3, we obtain

$$-\ln \mathcal{D}' \leq -\ln(\mathcal{V}^\dagger \mathcal{D} \mathcal{V}). \tag{B.29}$$

By combining inequalities (B.26) and (B.29), we finally obtain inequality (B.24), which implies the monotonicity (64). We note that the assumption of the complete positivity has been used only for the proof of the Schwarz's operator inequality (B.12), in which the assumption of the 2-positivity is in fact enough.

Acknowledgments

The author thanks to Yu Watanabe, Hal Tasaki, and Kouki Nakata for valuable comments. This work was supported by the Grant-in-Aid for Research Activity Start-up (KAKENHI 11025807).

References

1. A. Wehrl, *Rev. Mod. Phys.* **50**, 221 (1978).
2. M. Ohya and D. Petz, *"Quantum entropy and its use"* (Springer, Berlin, 2nd edition, 2004).
3. H. Umegaki, *Ködai Math. Sem. Rep.* **14**, 59 (1962).
4. S. Kullback and R. A. Leibler, *Ann. Math. Stat.* **22**, 79 (1951).
5. O. E. Lamford and D. W. Robinson, *Math. Phys.* **9**, 1120 (1968).
6. H. Araki and E. Lieb, *Commun. Math. Phys.* **18**, 160 (1970).
7. G. Lindblad, *Commun. Math. Phys.* **39**, 111 (1974).
8. G. Lindblad, *Commun. Math. Phys.* **40**, 147 (1975).
9. A. Uhlmann, *Commun. Math. Phys.* **54**, 21 (1977).
10. E. H. Lieb, *Advances in Mathematics* **11**, 267 (1973).
11. E. H. Lieb and M. B. Ruskai, *Phys. Rev. Lett.* **30**, 434 (1973).
12. E. H. Lieb and M. B. Ruskai, *J. Math. Phys.* **14**, 1938 (1973).
13. M. Ohya, *Rep. Math. Phys.* **27**, 19 (1989).
14. M. A. Nielsen and I. L. Chuang, *"Quantum Computation and Quantum Information"* (Cambridge University Press, Cambridge, 2000).
15. V. Vedral, *Rev. Mod. Phys.* **74**, 197 (2002).
16. M. Hayashi, *"Quantum Information: An Introduction"* (Springer-Verlag, Berlin, 2006).
17. D. Petz, *"Quantum Information Theory and Quantum Statistics"* (Springer, Berlin, 2008).
18. H. B. Callen, *"Thermodynamics and an Introduction to Thermostatistics, 2nd Edition,"* (John Wiley and Sons, New York, 1985).
19. D. J. Evans, E. G. D. Cohen, and G. P. Morriss, *Phys. Rev. Lett.* **71**, 2401 (1993).

20. G. Gallavotti, and E. G. D. Cohen, *Phys. Rev. Lett.* **74**, 2694 (1995).
21. C. Jarzynski, *Phys. Rev. Lett.* **78**, 2690 (1997).
22. G. E. Crooks, *Phys. Rev. E* **60**, 2721 (1999).
23. C. Maes, *J. Stat. Phys.* **95**, 367 (1999).
24. C. Jarzynski, *J. Stat. Phys.* **98**, 77 (2000).
25. U. Seifert, *Phys. Rev. Lett.* **95**, 040602 (2005).
26. R. Kawai, J. M. R. Parrondo, and C. Van den Broeck, *Phys. Rev. Lett.* **98**, 080602 (2007).
27. A. Gomez-Marin, J. M. R. Parrondo, and C. Van den Broeck, *Phys. Rev. E* **78**, 011107 (2008).
28. J. M. R. Parrondo, C. Van den Broeck, and R. Kawai, *New J. Phys.* **11**, 073008 (2009).
29. J. Kurchan, arXiv:cond-mat/0007360 (2000).
30. H. Tasaki, arXiv:cond-mat/0009244 (2000).
31. S. Yukawa, *J. Phys. Soc. Jpn.* **69**, (2000).
32. S. Mukamel, *Phys. Rev. Lett.* **90**, 170604 (2003).
33. C. Jarzynski and D. K. Wójcik, *Phys. Rev. Lett.* **92**, 230602 (2004).
34. W. De Roeck and C. Maes, *Phys. Rev. E* **69**, 026115 (2004).
35. T. Monnai, *Phys. Rev. E* **72**, 027102 (2005).
36. M. Esposito and S. Mukamel, *Phys. Rev. E* **73**, 046129 (2006).
37. P. Talkner and P. Hänggi, *J. Phys. A: Math. Theor.* **40**, F569 (2007).
38. P. Talkner, E. Lutz, and P. Hänggi, *Phys. Rev. E* **75**, 050102(R) (2007).
39. M. Esposito, U. Harbola, and S. Mukamel, *Phys. Rev. B* **75**, 155316 (2007).
40. K. Saito and A. Dhar, *Phys. Rev. Lett.* **99**, 180601 (2007).
41. D. Andrieux and P. Gaspard, *Phys. Rev. Lett.* **100**, 230404 (2008).
42. G. Huber, F. Schmidt-Kaler, S. Deffner, and E. Lutz, *Phys. Rev. Lett.* **101**, 070403 (2008).
43. K. Saito and Y. Utsumi, *Phys. Rev. B* **78**, 115429 (2008) .
44. Y. Utsumi and K. Saito, *Phys. Rev. B* **79**, 235311 (2009).
45. M. Esposito, U. Harbola, and S. Mukamel, *Rev. Mod. Phys.* **81**, 1665 (2009).
46. D. Andrieux, P. Gaspard, T. Monnai, and S. Tasaki, *New J. Phys.* **11**, 043014 (2009).
47. M. Campisi, P. Talkner, and P. Hänggi, *Phys. Rev. Lett.* **102**, 210401 (2009).
48. M. Campisi, P. Talkner and P. Hänggi, *Phys. Rev. Lett.* **105**, 140601 (2010).
49. S. Nakamura *et al.*, *Phys. Rev. Lett.* **104**, 080602 (2010).
50. M. Ohzeki, *Phys. Rev. Lett.* **105**, 050401 (2010).
51. M. Campisi, P. Hänggi, and P. Talkner, *Rev. Mod. Phys.* **83**, 771 (2011).
52. S. Deffner, M. Brunner, and E. Lutz, *Europhys. Lett.* **94**, 30001 (2011).
53. J. M. Horowitz, arXiv:1111.7199 (2011).
54. M. Esposito, K. Lindenberg, and C. Van den Broeck, *New J. Phys.* **12**, 013013 (2010).
55. H. Hasegawa, J. Ishikawa, K. Takara, and D.J. Driebe, *Phys. Lett. A* **374**, 1001 (2010).
56. M. B. Ruskai, *J. Math. Phys.* **43**, 4358 (2002).
57. M. B. Plenio, S. Virmani and P. Papadopoulos, *J. Phys. A: Math. Gen.* **33**, L193 (2000).

58. S. Yukawa, arXiv:cond-mat/0108421 (2001).

59. H. P. Breuer and F. Petruccione, *"The theory of open quantum systems,"* (Oxford, 2002).

60. T. Hatano and S.-I. Sasa, *Phys. Rev. Lett.* **86**, 3463 (2001).

61. T. Speck and U. Seifert, *J. Phys. A: Math. Gen.* **38** L581 (2005).

62. S. Sasa and H. Tasaki, *J. Stat. Phys.* **125**, 125 (2006).

63. M. Esposito, U. Harbola, and S. Mukamel, *Phys. Rev. E* **76**, 031132 (2007).

64. M. Esposito and C. Van den Broeck, *Phys. Rev. Lett.* **104**, 090601 (2010).

65. C. Pérez-Espigares, A. B. Kolton, and J. Kurchan, arXiv:1110.0967 (2011).

66. A. S. Holevo, *Probl. Inf. Transm.* **9**, 177 (1973).

67. H. P. Yuen and M. Ozawa, *Phys. Rev. Lett.* **70**, 363 (1993).

68. C. A. Fuchs and C. M. Caves, *Phys. Rev. Lett.* **73**, 3047 (1994).

69. H. J. Groenewold, *Int. J. Theor. Phys.* **4**, 327 (1971).

70. M. Ozawa, *J. Math. Phys.* **27**, 759 (1986).

71. F. Buscemi, M. Hayashi, and M. Horodecki, *Phys. Rev. Lett.* **100**, 210504 (2008).

72. S. Lloyd, *Phys. Rev. A* **39**, 5378 (1989).

73. S. Lloyd, *Phys. Rev. A* **56**, 3374 (1997).

74. M. A. Nielsen, C. M. Caves, B. Schumacher, and H. Barnum, *Proc. R. Soc. London A* **454**, 277 (1998).

75. W. H. Zurek, arXiv:quant-ph/0301076 (2003).

76. T. D. Kieu, *Phys. Rev. Lett.* **93**, 140403 (2004).

77. A.E. Allahverdyan, R. Balian, and Th.M. Nieuwenhuizen, *J. Mod. Optics* **51**, 2703 (2004).

78. H. T. Quan, Y. D. Wang, Y-x. Liu, C. P. Sun, and F. Nori, *Phys. Rev. Lett.* **97**, 180402 (2006).

79. T. Sagawa and M. Ueda, *Phys. Rev. Lett.* **100**, 080403 (2008).

80. K. Jacobs, *Phys. Rev. A* **80**, 012322 (2009).

81. S. W. Kim, T. Sagawa, S. De Liberato, and M. Ueda, *Phys. Rev. Lett.* **106**, 070401 (2011).

82. H. Dong, D. Z. Xu, C. Y. Cai, and C. P. Sun, *Phys. Rev. E* **83**, 061108 (2011).

83. Y. Morikuni and H. Tasaki, *J. Stat. Phys.* **143**, 1 (2011).

84. D. Abreu and U. Seifert, *Phys. Rev. Lett.* **108**, 030601 (2012).

85. S. Lahiri, S. Rana, and A. M. Jayannavar, *J. Phys. A: Math. Theor.* **45**, 065002 (2012).

86. Y. Lu and G. L. Long, *Phys. Rev. E* **85**, 011125 (2012).

87. M.-D. Choi, *Linear Algebra and Its Applications* **10**, 285 (1975).

88. A. Peres, *Phys. Rev. Lett.* **77**, 1413 (1996).

89. M. B. Plenio, *Phys. Rev. Lett.* **95**, 090503 (2005)

90. J. von Neumann, *Mathematische Grundlagen der Quantumechanik* (Springer, Berlin, 1932) [Eng. trans. R. T. Beyer, *Mathematical Foundations of Quantum Mechanics* (Prinston University Press, Princeton, 1955)].

91. E. B. Davies and J. T. Lewis, *Commun. Math. Phys.* **17**, 239 (1970).

92. W. F. Stinespring, *Proc. Amer. Math. Soc.* **6**, 211 (1955).

93. K. Kraus, *Ann. Phys.* **64**, 311 (1971).

94. M. Ozawa, *J. Math. Phys.* **25**, 79 (1984).
95. K. Koshino and A. Shimizu, *Phys. Rept.* **412**, 191 (2005).
96. C. Shannon, *Bell System Technical Journal* **27**, 379-423 and 623-656 (1948).
97. T. M. Cover and J. A. Thomas, *"Elements of Information Theory"* (John Wiley and Sons, New York, 1991).
98. D. Petz, *Rep. Math. Phys.* **23** 57 (1986).
99. M. A. Nielsen and D. Petz, *Quantum Inf. Comput.* **5**, 507 (2005).
100. D. Petz, *Rev. Math. Phys.* **15**, 79 (2003).
101. H. M. Wiseman and G. J. Milburn, *"Quantum Measurement and Control"* (Cambridge, UK: Cambridge University Press, 2010).
102. T. Sagawa and M. Ueda, *Phys. Rev. Lett.* **104**, 090602 (2010).
103. Y. Fujitani and H. Suzuki, *J. Phys. Soc. Jpn.* **79**, 104003 (2010).
104. J. M. Horowitz and S. Vaikuntanathan, *Phys. Rev. E* **82**, 061120 (2010).
105. S. Toyabe, T. Sagawa, M. Ueda, E. Muneyuki, and M. Sano, *Nature Physics* **6**, 988 (2010).
106. S. Ito and M. Sano, *Phys. Rev. E* **84**, 021123 (2011).
107. T. Sagawa and M. Ueda, *Phys. Rev. E* **85**, 021104 (2012).
108. L. Szilard, *Z. Phys.* **53**, 840 (1929).
109. D. Abreu and U. Seifert, *Europhys Lett.* **94**, 10001 (2011).
110. J. M. Horowitz and J. M. R. Parrondo, *Europhys Lett.* **95**, 10005 (2011).
111. G. Lindblad, *Commun. Math. Phys.* **48**, 119 (1976).
112. C. Gardiner and P. Zoller, *"Quantum Noise: A Handbook of Markovian and Non-markovian Quantum Stochastic Methods With Applications to Quantum Optics"* (Springer, Berlin, 3rd edition, 2004).
113. A. E. Allahverdyan and T.M. Nieuwenhuizen, *Phys. Rev. E* **64**, 0561171 (2001).
114. C. Horhammer and H. Buttner, *J. Stat. Phys.* **133**, 1161 (2008).
115. D. Jennings and T. Rudolph, *Phys. Rev. E* **81**, 061130 (2010).
116. L. Maccone, *Phys. Rev. Lett.* **103**, 080401 (2009).
117. D. Jennings and T. Rudolph, *Phys. Rev. Lett.* **104**, 148901 (2010).
118. J. von Neumann, *Z. Phys.* **57**, 30 (1929) [Eng. Trans. in arXiv:1003.2133].
119. J. M. Deutsch, *Phys. Rev. A* **43**, 2046 (1991).
120. M. Srednicki, *Phys. Rev. E* **50**, 888 (1994).
121. H. Tasaki, *Phys. Rev. Lett.* **80**, 1373 (1998).
122. S. Goldstein, J. L. Lebowitz1, R. Tumulka, and N. Zanghi, *Phys. Rev. Lett.* **96**, 050403 (2006).
123. S. Popescu, A. J. Short, and A. Winter, *Nature Physics* **2**, 754 (2006).
124. P. Reimann, *Phys. Rev. Lett.* **99**, 160404 (2007).
125. T. N. Ikeda, Y. Watanabe, and M. Ueda, *Phys. Rev. E* **84**, 021130 (2011).
126. S. Sugiura and A. Shimizu, arXiv:1112.0740 (2012).
127. T. Kinoshita, T. Wenger, and D. S. Weiss, *Nature* **440**, 900 (2006).
128. A. Lenard, *J. Stat. Phys.* **19**, 575 (1978).
129. W. Pusz and S. L. Woronowic, *Commun. Math. Phys.* **58**, 273 (1978).
130. H. Tasaki, arXiv:cond-mat/0009206 (2000).
131. M. Campisi, *Stud. Hist. Phil. Mod. Phys.* **39**, 181 (2008).
132. K. Sato, *J. Phys. Soc. Jpn.* **71**, 1065 (2002).

133. R. Marathe and J. M. R. Parrondo, *Phys. Rev. Lett.* **104**, 245704 (2010).

134. S. Vaikuntanathan and C. Jarzynski, *Phys. Rev. E* **83**, 061120 (2011).

135. P. Talkner, P. Hänggi, and M. Morillo, *Phys. Rev. E* **77**, 051131 (2008).

136. H. E. D. Scovil and E. O. Schulz-DuBois, *Phys. Rev. Lett.* **2**, 262 (1959).

137. J. E. Geusic, E. O. Schulz-DuBois, and H. E. D. Scovil, *Phys. Rev.* **156**, 343 (1967).

138. C. M. Bender, D. C. Brody, and B. K. Meister, *J. Phys. A: Math. Gen.* **33**, 4427 (2000).

139. M. O. Scully, *Phys. Rev. Lett.* **87**, 220601 (2001).

140. M. O. Scully *et al.*, *Science* **299**, 862 (2003).

141. H. T. Quan, Yu-xi Liu, C. P. Sun, and F. Nori, *Phys. Rev. E* **76**, 031105 (2007).

142. S. De Liberato and M. Ueda, *Phys. Rev. E* **84**, 051122 (2011).

143. J. C. Maxwell, *"Theory of Heat,"* (Appleton, London, 1871).

144. *"Maxwell's demon 2: Entropy, Classical and Quantum Information, Computing"*, H. S. Leff and A. F. Rex (eds.), (Princeton University Press, New Jersey, 2003).

145. K. Maruyama, F. Nori, and V. Vedral, *Rev. Mod. Phys.* **81**, 1 (2009).

146. O. J. E. Maroney, "Information Processing and Thermodynamic Entropy," The Stanford Encyclopedia of Philosophy (Fall 2009 Edition), Edward N. Zalta (ed.).

147. T. Sagawa and M Ueda, *"Information Thermodynamics: Maxwell's Demon in Nonequilibrium Dynamics,"* arXiv:1111.5769 (2011); to appear in R. Klages, W. Just, C. and Jarzynski (Eds.), *"Nonequilibrium Statistical Physics of Small Systems: Fluctuation Relations and Beyond"* (Wiley-VCH, Weinheim, 2012).

148. O. J. E. Maroney, *Phys. Rev. E* **79**, 031105 (2009).

149. T. Sagawa and M. Ueda, *Phys. Rev. Lett.* **102**, 250602 (2009); **106**, 189901(E) (2011).

150. R. Bhatia, *"Matrix analysis"* (Springer-Verlag, New York, 1997).

151. T. Ando, *"Topics of operator inequalities,"* Lecture notes, Hokkaido Univ. (Sapporo, 1978).

152. C. Davis, *Proc. Am. Math. Soc.* **8**, 42 (1957).

153. F. Hansen and G. K. Pedersen, *Math. Ann.* **258**, 229 (1982).

154. F. Hansen and G. K. Pedersen, *Bull. London Math. Soc.* **35**, 553 (2003).

155. R. V. Kadison, *Ann. of Math.* **56**, 494 (1952).

156. M.-D. Choi, *Illinois J. Math.* **18**, 565 (1974).